FLORA OF TROPICAL EAST AFRICA

SANTALACEAE

R.M. POLHILL[1]

Herbs, shrubs or trees, hemiparasitic. Leaves often alternate, sometimes opposite, petiolate or sessile, simple, entire, sometimes reduced to scales, exstipulate. Flowers in various sorts of essentially cymose inflorescences, often with a small dichasium axillary to each bract, small, often greenish, hermaphrodite or unisexual (then plants monoecious or dioecious), regular, monochlamydeous, the lobes distinct (in tropical African genera), forming a valvate 3–5-lobed fleshy cup or tube. Stamens as many as and opposite the lobes, inserted at or below their base; anthers (in tropical African genera) with 2 parallel thecae opening by longitudinal slits. Disk epigynous, intrastaminal, lobed, often lining at least the lower part of the perianth tube. Ovary (in tropical African genera) inferior, unilocular; placenta erect, free-central with 2–3 pendulous ovules; style simple, cylindric or sometimes nearly wanting; stigma terminal, capitate or 2–3(–5)-lobed. Fruit indehiscent, dry or fleshy (nut or drupe). Seed solitary; testa obsolete; cotyledons surrounded by copious fleshy, oily or starchy endosperm.

A family of about 35 genera and at least 400 species, nearly cosmopolitan, but commonest in tropical and subtropical regions. By far the largest genus is *Thesium*, which accounts for more than half the species and is native chiefly to Africa and the Mediterranean region.

1. Young twigs and peduncles compressed, elliptic in section; leaves broad, penninerved, minutely white-dotted on both surfaces . . . . 3. **Osyris**
   Young twigs and peduncles not compressed though often angled, ribbed or winged; leaves often very narrow or reduced to scales, if broad, then distinctly 3-nerved, not white-dotted 2
2. Leaves usually reduced to scales or very narrow (up to 2 mm broad), rarely up to ± 6 mm broad and then sessile; bracts adnate to peduncle (where present) . . . . 1. **Thesium**
   Leaves broad, distinctly petiolate; bracts free from peduncles 2. **Osyridicarpos**

Santalum album L., Sandalwood, native of Asia, has been grown as a street tree and hedge in Dar es Salaam, e.g. State House, 6 Sept. 1972, *Ruffo* 492! and Yacht Club, 5 Dec. 1947, *Wigg* in *F.H.* 2273!, and as a plantation tree in the Pugu Hills, June 1954, *Semsei* 1754! & 17 Jan. 1998, *Abeid* 206!, and at Morogoro, Mar. 1948, *Wigg* in *F.H.* 2286!; also in the W Usambara Mts and at Kilosa, see F.D.-O.A. 2: 144 (1932) and T.T.C.L.: 351 (1949). Evergreen shrub or small tree; leaves opposite, petiolate, often glaucous beneath, elliptic to ovate-elliptic, 4–9 × 1.5–3.5 cm, pinnately nerved; flowers in thyrses, 1–3 per bract; perianth pale green turning purplish inside,

[1] This text is largely adapted, in part verbatim, from the lucid and scholarly account by Dr Olive Hilliard for Flora Zambesiaca (ined.). I am most grateful to Mr Quentin Luke for assessing the material in the East African Herbarium (EA) and selecting a loan, to Dr Robert Vogt for images and information from the Botanical Museum Berlin-Dahlem (B) and to Mr Roy Gereau for information and selecting a loan from the Missouri Botanical Garden (MO).

the lobes 4(–5), 2 mm long, bearing a small stamen in a tuft of hairs; fruit a drupe, reddish purple turning black, subglobose, 1–1.5 cm diameter. The wood is highly valued for its fragrance used in perfumes and cosmetics. The essence is distilled from the heartwood. The wood is one of the finest for carving. *Senna siamea* is recommended as a host.

1. **THESIUM**

L., Sp. Pl.: 207 (1753) & Gen. Pl., ed. 5: 97 (1754)

Shrubs or herbs. Leaves sessile, alternate, oblong, linear or reduced to scales, 3-nerved. Flowers hermaphrodite, axillary and terminal, solitary or in cymules; bracts generally conspicuous, often adnate to the pedicel or peduncle; bracteoles 2–several, persistent. Perianth leathery; tube campanulate to cylindrical, often with evident glands below the sinuses of the lobes, sometimes tooth-like; lobes (3–)5, valvate, glabrous or variously fringed or bearded, the tips often hooded. Stamens as many as the lobes, inserted on or at the base of the lobes; anthers in East African species attached to the lobes by a tuft of hairs. Disk developed to varying degrees, fleshy, lobed, lining the perianth tube. Ovary with 3 ovules pendent from apex of a free-central placenta, the placenta straight or twisted; style usually evident, simple. Fruit usually shortly stipitate, indehiscent, globose or ovoid, usually bony, 10-ribbed and ± reticulate between the ribs, or less often fleshy, crowned with the persistent perianth.

More than 300 species in Africa and Eurasia, about 175 in southern Africa (42 in the area of Flora Zambesiaca, 17 in East Africa, 8 in Ethiopia).

The genus is considered to be semiparasitic on a wide range of plants without evident host specificity, but no local studies have been reported. Some general information on the morphology, natural history and classification of the genus is to be found in a discussion of the Eurasian species by Hendrych in Acta Univ. Carolinae, Biol. 1970, No. 4: 293–358 (1972). The small fruits are adapted for dispersal by ants, but again there appear to have been no local studies. Although a number of species are relatively widespread in East Africa, populations tend to be scattered and the genus is still fairly thinly represented in herbaria. Some localities do seem to support several species and it is possible that the development of an association with ants is a significant factor in the distribution. The genus has a reputation for being difficult taxonomically, partly due to the sparsity of material, and certainly the use of a good dissecting microscope is helpful to see structure in the small flowers, but the existence of recent accounts for the surrounding regions has now made the analysis of the East African species much easier. The species become much more numerous towards southern Africa and the excellent account by Dr Olive Hilliard for F.Z. (ined.) has been invaluable. The East African species can be allotted to 6 groups following the scheme she suggests. Groups 1–5 belong to subgen. *Thesium* (sect. *Imberbia* A.W.Hill) and Group 6 to subgen. *Frisea* (Endl.) Petermann (sect. *Barbata* A.W.Hill) in Hendrych's classification, which does not, however, include any significant analysis of the many African species considered by Arthur Hill in F.T.A. (1911), K.B. 1915: 10 (1915) and Fl. Cap. 5(2): 137–144 (1915). The bunch of hairs situated beyond the stamens in subgen. *Frisea* contains, at least in some Eurasian species, a thick, oily to resinous secretion, which flows when touched. This, taken with the small, dull leathery flowers bearing external glands and a fleshy internal disk, often arranged within bracts close to the main axes, rather suggests that some species at least may be pollinated as well as dispersed by ants or by small oil-seeking bees. There is often a contradiction on field notes that variously describe the flowers as yellow or white. It seems that they are usually yellow-green outside, white on the face of the lobes inside at anthesis, but then turn yellowish later. There are some species, however, that do seem to have flowers that are yellowish throughout their maturation.

Group 1. Shrub; stems slender, climbing or straggling, with ribs sharply raised and extending into midribs; leaves well developed; flowers in dichasial cymes; bracts usually adnate to peduncle; bracteoles 2; perianth glands inconspicuous, broadly ellipsoid; lobes subacute, only slightly hooded, glabrous; stamens at base of perianth lobes; filaments short; placenta twisted; fruit thinly fleshy, without raised veins. Species 1.

Group 2. Annual or perennial herbs; stems spreading to erect, with ribs broad and shallow from decurrent leaves or raised sharply and extending into midribs; leaves all well developed; flowers single in the bracts or cymose; bracts basally adnate to subtended stalk; bracteoles 2; perianth glands minute or hardly evident; lobes subacute, glabrous; stamens at base of perianth lobes; filaments short; placenta straight; fruits weakly to moderately ribbed and reticulate. Species 2, 3.

Group 3. Perennial herbs, with stems from conspicuous vegetative buds, strongly ribbed; leaves scale-like basally, well developed above; flowers solitary, sessile in each bract, spiciform; bracteoles 2; perianth glands ellipsoid; lobes fringed with long hairs; stamens on lower part of perianth lobes, with conspicuous filaments; placenta straight; fruit with moderately raised ribs and veins. Species 4.

Group 4. Annual or perennial herbs with a narrow rootstock, sometimes rhizomatous; stems erect, with sharply raised ribs to narrowly winged; leaves reduced to scales, but annual species with one pair of basal leaves; flowers 1(–3), sessile in the bracts, spiciform or crowded on short shoots; bracteoles 2(–4); perianth glands tooth-like, rounded swellings, elongate or obscure; lobes subacute and glabrous to distinctly hooded and ciliate; stamens inserted at base of perianth lobes or a little above, the filaments conspicuous or less often almost hidden by anthers; placenta straight; fruit ribbed and vein reticulum raised to varying degrees. Species 5–10.

Group 5. Perennial herbs with a broad rootstock crowned with vegetative buds; stems erect with shallow ribs from decurrent scales; leaves usually all reduced to scales, but sometimes developed leaves intermixed in *T. fastigiatum*; flowers at ends of naked or scaly peduncles, each with an involuce of ± 5 bracts; perianth glands not very evident; lobes slightly hooded, glabrous to erose; stamens at base of perianth lobes, ± hidden by anthers; placenta straight; fruits strongly ribbed and reticulately veined. Species 11–13.

Group 6. Perennal herbs with a narrow rootstock, usually rhizomatous; stems prostrate to erect, with broad, shallow to sharply raised ribs; leaves mostly well developed, the basal ones ± reduced to scales; flowers 1(–3) in axils of bracts, racemose or cymose; bracteoles 2; perianth glands ellipsoid or inconspicuous; lobes white-bearded at the apex; stamens at base of perianth lobes, visible or hidden by anthers; placenta straight; fruits ribbed and reticulate to varying degrees or (*T. radicans*) red and fleshy. Species 14–17.

1. Leaves all well developed, except generally the basal ones 2
   Leaves all or mostly reduced to scales (*T. brachyanthum* and *T. subaphyllum* have 2 linear basal leaves; *T. fastigiatum* sometimes has some linear leaves intermixed) 9
2. Stems generally scrambling, with pendent inflorescences (fig. 2); leaves linear to oblong, ± 10–60 × (1–)2–6 mm; fruit orange to red, with thinly fleshy epicarp, the bony endocarp 5-veined but not ribbed, irregularly ridged and wrinkled . . . 1. *T. triflorum*
   Stems prostrate, tufted or shortly erect; leaves up to ± 2 mm wide; fruits either hard and ribbed or orange to red and fleshy 3
3. Perianth lobes glabrous inside or hairy to erose along the margins (fig. 1/6, 8), but not bearded at the apex inside 4
   Perianth lobes white-bearded at the apex inside (fig. 1/16) 6
4. Stems from a taproot, not arising from vegetative buds, much branched from base, tufted or prostrate, densely leafy, broadly ribbed by decurrent leaves, distinctly channelled between the ribs; perianth lobes (3–)4(–5), 0.5–1 mm long; fruits 1.5–2 mm long . . . . 2. *T. kilimandscharicum*
   Stems from a woody rootstock, emerging from conspicuous, cone-like vegetative buds, erect, leafy upwards, with sharply raised ribs decurrent from the midribs of leaves and bracts; perianth lobes 5, 1–4.5 mm long; fruits 2.2–3.5 mm long 5

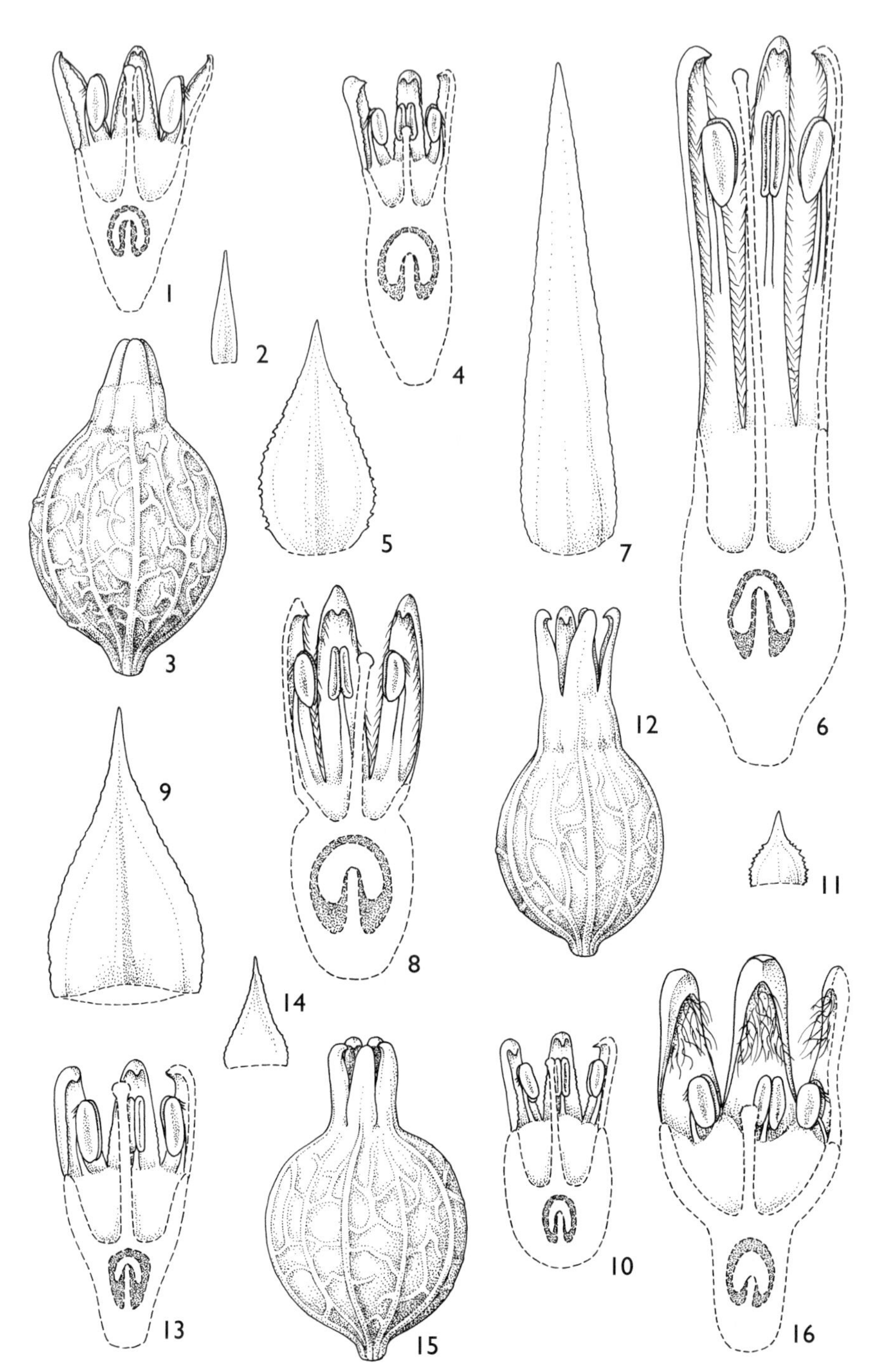

FIG. 1. Flowers, in longitudinal section, bracts and fruits, all × 10, of *Thesium* species. **1–3**, *T. subaphyllum*; **4**, **5**, *T. brachyanthum*; **6**, **7**, *T. ussanguense*; **8**, **9**, *T. fimbriatum*; **10–12**, *T. fastigiatum*; **13–15**, *T. unyikense*; **16**, *T. goetzeanum*. 1–3, from *Drummond & Rutherford-Smith* 7208; 4, 5, from *Sanane* 780; 6, 7, from *Torre & Pereira* 12809; 8, 9, from *Fanshawe* 4780; 10–12, from *Brummitt* 18339; 13–15, from *Pawek* 1395; 16, from *Fanshawe* 8951. Drawn by Mary Bates. Reproduced with permission from Flora Zambesiaca.

5. Vegetative buds 1–2 mm long, with scales 0.2–0.8 mm long; perianth lobes 1–1.2 mm long, minutely ciliolate; flowers cymose at the apex of the stems (fig. 3) . . . . 3. *T. panganense*
   Vegetative buds 5–10 mm long, with scales 2–6 mm long; perianth lobes 2–4.5 mm long, densely fringed with long white hairs (fig. 1/6); flowers in long spikes from near the base of the stems upwards . . . . 4. *T. ussanguense*
6. Plants forming a mat, spreading by slender rhizomes, the flowering stems or branches erect for only 3–10 cm, the leafy branches long and trailing; perianth lobes 0.6–0.8 mm long; fruits orange to red, fleshy . . . . 14. *T. radicans*
   Plants erect; perianth lobes 0.8–2 mm long; fruits buff to reddish brown, dry 7
7. Leaves glaucous (drying mauvish grey), with midrib raised at least basally and decurrent as a ± sharp ridge; inflorescence normally cymose in part, the flowers there in groups of 2–3, other flowers solitary; fruits 3–3.5 mm long, with ribs generally more prominent than the reticulation between (at least before final maturation) . . . . 17. *T. goetzeanum*
   Plants greenish; ribs shallow and broad, formed from decurrent leaves; flowers solitary in each bract, arranged in racemes; fruits with strong ribs and reticulate venation 8
8. Leaves flat, green, ± straight, with a cartilaginous margin and prominent tip, the midrib evident; flowers up to 5(–7) per raceme, the bracts sometimes short but often leafy and 12 mm or more long; perianth lobes 0.8–1 mm long; anthers 0.25–0.3 mm long; fruits 2.2–3 mm long; rootstock without narrow rhizomes but sometimes thickened and shortly branched at the crown . . . . 15. *T. schweinfurthii*
   Leaves slightly thickened, yellow-green to grey-green, often slightly curved upwards, the edges rounded, the midrib obscured; flowers mostly 7–14 per raceme, the bracts ± 5–17 mm long, often mostly less than 12 mm long; perianth lobes 1.2–1.7 mm long; anthers 0.5–0.8 mm long; fruits 3–4 mm long; rootstock small, with slender rhizomes 1–3(–4) mm thick . . . . 16. *T. mukense*
9. Perennial with rootstock often broader than deep and crowned with numerous vegetative buds that eventually grow out into stems; ribs on stems broad and low; flowers terminating naked or scaly peduncles, each flower surrounded by an involucre of ± 5 small bracts 10
   Annual or perennial with a taproot, which is sometimes swollen and persistent, sometimes with rhizomes, but vegetative buds not very evident; stems with sharply raised ribs or wings extending into the midribs of scale leaves or bracts; flowers sessile in the axil of a bract and 2(–4) bracteoles, arranged in spikes, which are sometimes condensed on short shoots 12

10. Margins and back of bracts, scale leaves and generally stems, at least near the base, conspicuously covered with stiff spreading hairs . . . . 13. *T. thamnus*
    Bracts only minutely ciliolate, the backs and other parts glabrous 11
11. Flowers mostly on scaly peduncles (branches usually subtended by scale leaves but occasionally also by linear leaves), becoming cymose at the tip and sometimes growing out as sterile scaly tails, a few flowers sometimes also on short naked peduncles . . 11. *T. fastigiatum*
    Flowers initially terminating peduncles with 0–1 scales below the bracts, but then overtopped by branches arising immediately or shortly below the flowers, and the pattern repeated . . . . 12. *T. unyikense*
12. Annual with a pair of ephemeral linear leaves near the base of the stem (when fallen scars still evident); perianth lobes triangular, 0.6–1.2 mm long, glabrous on the margins (fig. 1/1, 4) 13
    Perennial, sometimes flowering in first year of growth, without paired basal leaves; perianth lobes linear-triangular to oblong or subspathulate, 1.5–3.6 mm long, glabrous to long ciliate (fig. 1/8) 14
13. Stem with wings up to 1 mm wide; bracts linear-lanceolate, 1–1.5 × 0.4–0.6 mm (fig. 1/2); fruit ellipsoid-globose to globose, with the reticulation rather more raised than the ribs (fig. 1/3) . . . . 5. *T. subaphyllum*
    Stem ribbed but not winged; bracts lanceolate, 1.5–3.5 × 1.7–2 mm (fig. 1/5); fruit obovoid, strongly ribbed and reticulate . . . . 6. *T. brachyanthum*
14. Perianth tube 1.5–2.5 mm long, the lobes 1.2–1.5 times as long, the tube with elongate glands (not always very obvious) from near the base to near the sinuses between the lobes; fruit strongly ribbed and reticulate; flowers all well-spaced in lax spikes . . . . 7. *T. microphyllum*
    Perianth tube 0.5–1.2 mm long, the lobes 1.8–3.5 times as long, the tube with small teeth at the sinuses of the lobes or with small rounded swellings below the sinuses; fruit strongly ribbed but reticulation slight; flowers sometimes all or partly in lax spikes, but usually mostly several–10 crowded on condensed short shoots or paired to crowded apically 15
15. Perianth with glabrous lobes and small swellings (sometimes obscure) below sinuses between the lobes 8. *T. stuhlmannii*
    Perianth lobes fringed with long hairs and with small tooth-like glands between them 16
16. Perianth lobes 1.5–2.7 mm long, subobtuse to acute, the hood 0.1–0.25(–4) mm long (fig. 1/8) . . . . 9. *T. fimbriatum*
    Perianth lobes 3–3.6 mm long, very acute, the hood 0.6–0.8 mm long . . . . 10. *T. schliebenii*

1. **Thesium triflorum** *L.f.*, Suppl. Pl.: 162 (1782); Thunberg, Prodr. Pl. Cap.: 46 (1794) & Fl. Cap., ed. Schultes: 211 (1823); Hill in K.B. 1915: 12, fig. 2 (1915) & in F.C. 5(2): 155 (1915); N.E. Brown in Burtt Davy, Fl. Pl. & Ferns Trans.: 462 (1932); Stauffer in Viert. Nat. Ges. Zürich 106: 400, fig. 5 (1961); Brummitt in K.B. 31: 176 (1976); White, Dowsett-Lemaire & Chapman, Evergreen For. Fl. Malawi: 520 (2001). Type: South Africa, Cape, *Herb. Thunberg* 6050 (UPS, holo.; microfiche!)

Small shrub scrambling 1.5–6 m or straggling if unsupported; stem slender and rough with leaf scars below, much branched above, leafy and grading into extensive leafy panicles; branches distinctly ribbed by raised vascular strands terminating in midribs of leaves and bracts. Leaves all well developed and well spaced, grading into bracts, linear to oblong, sometimes slightly falcate, very variable in size, mostly 10–60 × (1–)2–6 mm, smaller on young shoots, acute to obtuse, base slightly narrowed, margins cartilaginous, smooth to minutely scabridulous, 3-nerved, veins slightly raised on both surfaces, smooth to scabridulous. Flowers subsessile in dichasial cymes composed of 3–many flowers, these further arranged in panicles, in scandent plants pendent with the leaves and bracts reflexed to stand erect in the hanging position; bracts mostly 10–40 × 1.2–6 mm, leaf-like but sometimes reduced at apex of panicle, adnate ± 1–3 mm to base of peduncle; bracteoles 2, ovate, ± 1–1.5 × 0.5–1.3 mm, acute, sometimes lowermost pair leaf-like and then up to ± 22 × 1 mm; primary peduncles ± 6–95 mm long, progressively smaller upwards. Perianth greenish yellow outside, yellow to cream or white inside; tube 0.7–1 mm long, external glands inconspicuous, broadly ellipsoid; lobes triangular, 1.2–1.4 mm long, subacute, slightly hooded, margins flat, glabrous. Stamens inserted at base of perianth lobes; filaments 0.3–0.5 mm long, almost hidden by anthers; anthers 0.3–0.5 mm long. Style 0.5–0.8 mm long; stigma reaching ± base of lobes; placenta twisted. Fruit orange to red when ripe, ellipsoid-globose to globose, 4–5 mm long, 3–5 mm in diameter, epicarp thin, fleshy, endocarp bony, 5 main veins visible, only slightly raised, some or all branching once at the base, intercostal areas irregularly reticulate; stipe up to 1 mm long. Fig. 2 (page 8).

TANZANIA. Lushoto District: Mazumbai Forest Reserve, 26 Apr. 1975, *Hepper & Field* 5152!; Morogoro District: Uluguru Mts, ridge E of Mwere R., 29 Dec. 1974, *Polhill & Wingfield* 4610!; Njombe District: about 13 km S of Njombe, 10 July 1956, *Milne-Redhead & Taylor* 11040!

DISTR. **T** 3, 6, 7; Zambia and Malawi (Nyika Plateau), then disjunctly to southern Mozambique and South Africa to the Eastern Cape Province (see Brummitt, loc. cit. (1976))

HAB. Edges of wet and dry montane forest, heath scrub; 1500–2700 m

SYN. *Osyridicarpos linearifolius* Engl. in E.J. 28: 385 (1900); Baker & Hill in F.T.A. 6(1): 432 (1911); F.D.-O.A. 2: 143 (1932); T.T.C.L.: 550 (1949). Lectotype, chosen by Stauffer (1961): Tanzania, Morogoro District, Uluguru Mts, Lukwangule Plateau, *Goetze* 310 (B!, lecto.)

O. linearifolius Engl. var. *goetzei* Engl. in E.J. 30: 305 (1901); Baker & Hill in F.T.A. 6(1): 433 (1911); F.D.-O.A. 2: 144 (1932); T.T.C.L.: 550 (1949). Type: Tanzania, Njombe District, Ukinga [Kinga], Pikurugwe Mt, *Goetze* 1253 (B!, holo.; BM!, EA, iso.)

Osyris angustifolia Baker in K.B. 1910: 238 (1910). Type: Mozambique, Inhambane, Aug. 1887, *Scott* s.n. (K!, holo.)

NOTE. Notably different from the other species of East African *Thesium*. Formerly it was sometimes confused with *Osyridicarpos*, but has narrower sessile leaves. The generic position was discussed by Stauffer, loc. cit. (1961).

2. **Thesium kilimandscharicum** *Engl.*, Hochgebirgsfl. Trop. Afr.: 200 (1892); Baker & Hill in F.T.A 6(1): 424 (1911); F.D.-O.A. 2: 142 (1932); Brenan in Mem. N.Y. Bot. Gard. 9: 65 (1954); A.V.P.: 75 (1957); A.G. Miller in Fl. Eth. & Erit. 3: 379, fig. 116.1.8–10 (1989); U.K.W.F., ed. 2: 159 (1994). Type: Tanzania, Moshi District, Mue R. [Muëbach], *H. Meyer* 248 (B!, holo.)

Annual or short-lived perennial, often rather yellowish green; stems numerous from the crown of the taproot, prostrate to tufted, 9–40 cm long, divaricately well-branched from the base, densely leafy, broadly ribbed by the decurrent leaves, distinctly channelled between the ribs. Leaves ascending-recurved to spreading, linear, main ones mostly 5–20 × 0.8–1 mm, apiculate, margins smooth, ± rounded on the back. Flowers solitary in axils of bracts, 3–5 in short axillary and terminal racemes; bracts adnate to very short pedicel, leaf-like, mostly 2.5–6 × 0.8–1 mm; bracteoles 2, similar to bracts but smaller; pedicels up to 0.75 mm long. Perianth yellow-green, white or yellowish inside; tube very short, inside the perianth tube

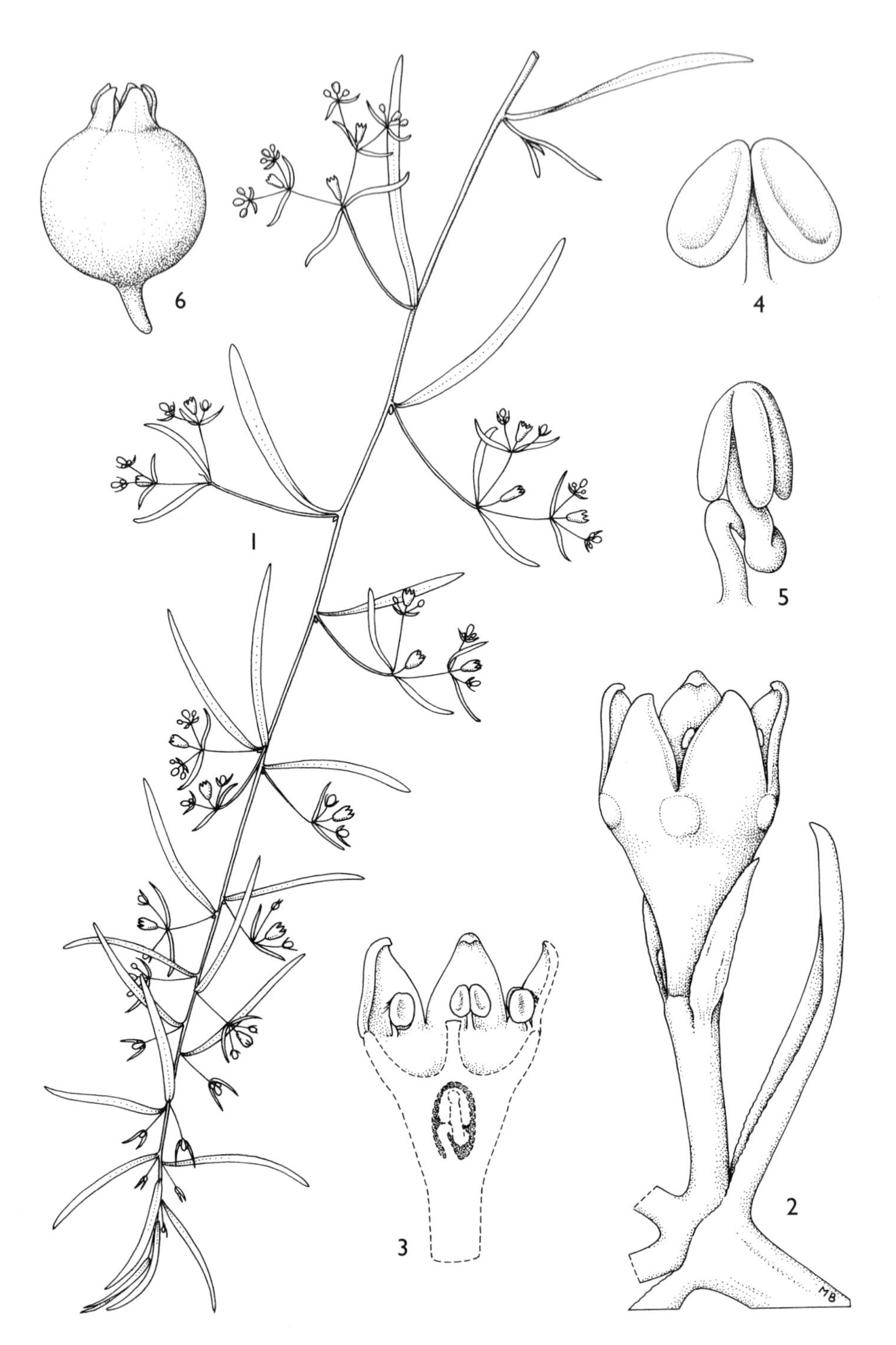

FIG. 2. *THESIUM TRIFLORUM* — **1,** part of pendent flowering branch, × 1; **2**, part of cyme, × 10; **3**, flower in longitudinal section, × 10; **4**, stamen, × 30; **5**, placenta and ovules, × 30; **6**, fruit, × 6. 1, from *Fanshawe* 7341; 2–6, from *Robson* 642. Drawn by Mary Bates. Reproduced with permission from Flora Zambesiaca.

completely hidden by the thick disk, external glands minute, almost globose; lobes (3–)4(–5), triangular, 0.5–1 mm long, subacute, glabrous. Stamens inserted at base of perianth lobes; filaments visible, 0.1–0.3 mm long; anthers 0.2–0.3 mm long. Style 0.2–0.3(–0.7) mm long; stigma reaching top of anthers or occasionally higher. Fruit light reddish brown, ellipsoid-globose, 1.5–2 mm long, 1.25–1.8 mm in diameter, weakly ribbed and reticulate; stipe pale, 0.5–0.75 mm long.

KENYA. Trans-Nzoia District: Elgon, Nov. 1940, *M. Jex-Blake* in *Bally* 1297!; Naivasha District: Aberdare Mts, S slope above S Kinangop Forest Station, 20 Sept. 1967, *Hedberg* 4339!; Meru District: NE Mt Kenya, Rotundu, 25 Sept. 1997, *Luke* 4776!

TANZANIA. Moshi District: Kilimanjaro, below Horombo [Peter's] Hut, 20 Jan. 1955, *Verdcourt* 1239!; Morogoro District: Uluguru Mts, Lukwangule Plateau, 19 Sept. 1970, *Thulin & Mhoro* 1010!; Njombe District: near Kikondo village on road to Kitulo, 15 Nov. 1982, *Magogo* 2308!

DISTR. **K** 2–4; **T** 2, 4, 6, 7; Ethiopia and Malawi (Nyika Plateau)

HAB. Montane grassland, heath scrub and afroalpine zone, on peaty and stony ground; 2200–4200 m

SYN. *T. kilimandscharicum* Engl. forma *decumbens* Engl., P.O.A. C: 168 (1895); F.D.-O.A. 2: 142 (1932); Chiovenda, Racc. Bot. Consol. Kenya: 110 (1935), *nom. superfl.* Type as for species

T. kilimandscharicum Engl. forma *erectum* Engl., P.O.A. C: 168 (1895); F.D.-O.A.: 142 (1932). Type: Tanzania, Kilimanjaro, Kibosho, Kifinika Crater, *Volkens* 1871 (B!, holo.)

T. ulugurense Engl. in E.J. 28: 385 (1900); Baker & Hill in F.T.A. 6(1): 423 (1911); F.D.-O.A. 2: 142 (1932); Vollesen in Opera Bot. 59: 63 (1980). Types: Tanzania, Morogoro District, Central Uluguru Mts, *Stuhlmann* 9206 (B†, syn.; K!, isoyn.) & Lukwangule Plateau, *Goetze* 256 (B, syn., seen by Hilliard in 1989)

T. rungwense Engl. in E.J. 30: 307 (1904). Type: Tanzania, Mt Rungwe, top crater, *Goetze* 1156 (B†, holo.; BM!, BR, K!, P, iso.)

NOTE. *T. kilimandscharicum* is easily recognised by its habit and (3–)4(–5)-lobed glabrous perianth, the tube so short that it is completely hidden inside by the thick disk, and the generally very short style, seldom more than 3 mm long.

3. **Thesium panganense** *Polhill*, **sp. nov**. *T. pallide* A.DC. valde affine sed tubo perianthii fere duplo longiore, 0.8–1 mm haud 0.5 mm, stylo manifeste longiore, 0.8–1 mm longo haud 0.25–0.4 mm, differt. Typus: Tanzania, Pangani District, between Pangani and Msubugwe Forest, *Milne-Redhead & Taylor* 7314 (K!, holo.; EA, iso.)

Perennial herb, the taproot thickened and forked at the apex to form a rootstock, and covered with small cone-like vegetative buds bearing scales 0.2–0.8 mm long, also producing rhizomes bearing further buds; stems dull green, numerous, erect, forming tufts, 13–20 cm tall, with scale leaves at the base grading into leaves, then soon producing ascending branches, the first ones sterile, the subsequent ones fertile above; ribs strongly raised and running into the midribs of leaves and bracts, with a slight tendency to lesser raised ribs running into the edges of the leaves (often more apparent in the inflorescences). Leaves: scales ovate to spathulate, 1–3 mm long; foliage leaves dull green, ascending, linear, 5–20 × 0.5–1 mm, acute and apiculate, minutely scabridulous along margins, the midrib slightly raised. Flowers 3–many in dichasial cymes, the upper part of the plant conspicuously floriferous, rather lax to slightly crowded terminally; primary peduncles 5–15 mm long, shorter upwards; bracts basally adnate to peduncles for 1–2 mm, leafy, linear, 5–12 × 0.4–0.8 mm; bracteoles similar to bracts but smaller. Perianth buff-coloured; tube 0.8–1 mm long, with glands no more than slight swellings below sinuses of lobes; lobes triangular, acute, 1–1.2 × 0.6–0.7 mm, minutely ciliolate. Stamens inserted at base of perianth lobes; filaments 0.2–0.3 mm long; anthers 0.3–0.4 mm long. Style 0.8–1 mm long; stigma reaching top of perianth tube or slightly above. Fruit reddish brown, ovoid-globose, 2.2–3 mm long, 2–2.5 mm in diameter, ribbed and with moderately raised reticulation between. Fig. 3 (page 10).

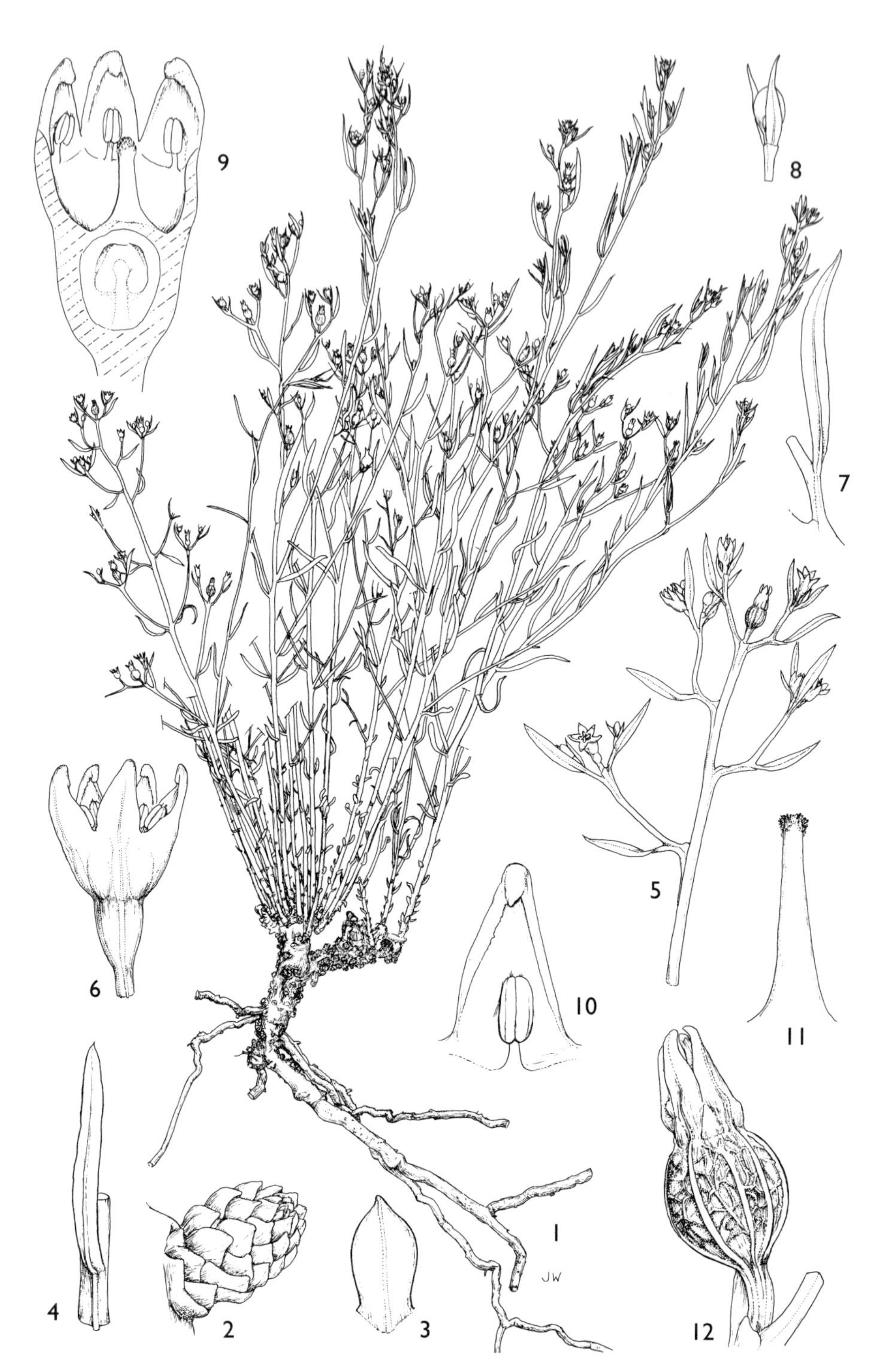

FIG. 3. *THESIUM PANGANENSE* — **1**, habit, × $^{2}/_{3}$; **2**, vegetative bud, × 10; **3**, scale leaf, × 10; **4**, leaf, showing insertion on stem, × 4; **5**, part of inflorescence, × 2; **6**, flower, × 10; **7**, bract, × 4; **8**, bracteoles and subtended flower bud, × 4; **9**, flower, longitudinal section, × 12; **10**, stamen attached to perianth-lobe, × 24; **11**, style, × 32; **12**, fruit, × 8. All from *Milne-Redhead & Taylor* 7314. Drawn by Juliet Williamson.

TANZANIA. Pangani District: between Pangani and Msubugwe Forest, 17 Nov. 1955, *Milne-Redhead & Taylor* 7314!
DISTR. **T** 3; known only from the type gathering
HAB. Scattered tree grassland after burning; ± 100 m

NOTE. Closely related to *T. pallidum* A.DC. (*T. floribundum* A.W.Hill), which is widespread in South Africa from the Limpopo Province to the Eastern Cape Province, Swaziland and Lesotho, but that species has a notably short cupular perianth tube and short style. *T. cymosum* A.W.Hill (*T. fenarium* A.W.Hill), from the highlands of southern Malawi and the Zimbabwe-Mozambique border mountains, is also similar vegetatively, but, despite the epithet, the single or 3-flowered cymules are arranged racemosely and the flowers are often all solitary except for the terminal cymes. In East Africa, *T. triflorum* is the only other species with dichasial cymes, each flower developing a bud in the axils of the bracteoles.

4. **Thesium ussanguense** *Engl.* in E.J. 30: 305 (1901); Baker & Hill in F.T.A. 6(1): 425 (1911); F.D.-O.A.: 142 (1932); U.K.W.F., ed. 2: 159 (1994). Type: Tanzania, Njombe District, Ussangu, Lipanga [Lipanye] Mts, *Goetze* 1264 (B, holo., seen by Hilliard in 1989; K!, iso.)

Perennial herb; rootstock thickened at the crown, up 4 cm across, often 2–several-fid and with short rhizomes, covered with very many conspicuous vegetative buds 5–10 mm long bearing straw-coloured scarious scales 2–6 mm long; flowering stems elongating from the buds, numerous, erect, forming tufts, 10–30 cm tall, simple to sparingly branched, the branches virgate, sharply ascending, closely leafy, often nearly the whole stem floriferous; ribs strongly raised and terminating in midrib of leaves and bracts. Leaves at extreme base of stem reduced to scales, straw-coloured, lanceolate-subulate, ± 4–8 × 1–2 mm; developed leaves green or grey-green, sharply ascending, lower ones often incurved, linear-lanceolate to linear, 5–15 × 1–1.6 mm, acuminate, apiculate, margins cartilaginous, pale, sometimes minutely ciliolate near base, outer surface keeled by strongly raised midrib. Flowers solitary, sessile in the axil of each bract, (10–)20–70 in long crowded spikes often with a long leafy sterile tip; bracts leaf-like, ± 4.5–11 × 0.8–2.4 mm; bracteoles 2, similar to bracts but often smaller, particularly narrower. Perianth pale greenish or creamy yellow outside, white inside; tube 1–2 mm long, external glands conspicuous, ellipsoid; lobes oblong, 2–4.5 mm long, ± obtuse, margins inflexed, densely fringed with long white hairs. Stamens inserted on lower part of perianth lobes; filaments 1–2 mm long, conspicuous; anthers 0.6–0.8 mm long. Style 2.5–5.5 mm long; stigma nearly reaching top of perianth lobes. Fruit pallid, ovoid-globose or ellipsoid-globose, 3–3.5 mm long, 2.5–3 mm in diameter, ribs and reticulations moderately raised. Fig. 1/6, 7 (page 4).

UGANDA. Acholi District: Imatong Mts, Apr. 1938, *Eggeling* 3551!; Karamoja District: Mt Moroto, 3 Jan. 1937, *A.S. Thomas* 2156!
KENYA. W Suk District: W of Kapenguria and ± 2 km E of Kaisakat Camp, 16 Apr. 1975, *Friis & Hansen* 2508!; Trans-Nzoia District: Kitale, Mar. 1966, *Tweedie* 3259!; Machakos District: Chyulu Hills saddle, 18 Jan. 1997, *Luke* 4596!
TANZANIA. Mpanda District: Mahali Mts, Kabesi, 1 Sept. 1958, *Newbould & Jefford* 2103!; Njombe District: between Lupalilo and Matamba, 27 Sept. 1970, *Thulin & Mhoro* 1209!; Songea District: Matengo Hills, Miyao, 12 Nov. 1956, *Semsei* 2587!
DISTR. **U** 1; **K** 2–4, 4/6; **T** 1, 2, 4, 7, 8; Rwanda, Burundi, Zambia, Malawi, Mozambique, Zimbabwe
HAB. Grassland, often in rocky places, coming up after fires; 1100–2400 m

SYN. *T. scoparium* Peter, F.D.-O.A. 2: 142, Anh. 12, t. 16/1 (1932), e descr. Type: Tanzania, Tabora District, near Kombe, *Peter* 35765 (B†, holo.)
T. passerinoides Robyns & Lawalrée in B.J.B.B. 31: 520 (1961). Type: Burundi, Mosso, Kininya, *Michel & Reed* 1531 (BR, holo.)

NOTE. Easily recognised by its tufted, leafy, and often simple stems arising from very conspicuous, straw-coloured vegetative buds, floriferous nearly to the base, flowers many in long crowded spikes, long heavily fringed perianth lobes, and anthers carried up on long filaments.

5. **Thesium subaphyllum** *Engl.*, P.O.A. C: 168 (1895); Baker & Hill in F.T.A. 6(1): 421 (1911); F.D.-O.A. 2: 142 (1932); Lawalrée in B.J.B.B. 39: 179 (1969); Vollesen in Opera Bot. 59: 63 (1980); A.G. Miller in Fl. Eth. & Erit. 3: 379, fig. 116.1.14 (1989); U.K.W.F., ed. 2: 159 (1994). Type: Tanzania, Kilimanjaro, Himo Hill, *Volkens* 1712 (B, holo., seen by Hilliard in 1989)

Annual herb with a small taproot; stems green to grey-green, 1(–few) from a taproot, erect, 0.3–1 m tall, virgately branched above, nearly leafless; ribs expanding into flattened wings up to ± 1 mm broad and terminating in the midribs of the bracts. Leaves: developed leaves only 2 near base of stem, opposite, linear, 20–30 × 1–1.4 mm; cauline leaves alternate, reduced to scales, closely appressed, lanceolate, 1–2 × 0.4–0.8 mm, acuminate, midrib strongly raised beneath. Flowers solitary in axils of bracts, sessile, 10–50 in lax spikes, laxly panicled; bracts linear-lanceolate, 1–1.5 × 0.4–0.6 mm, acuminate; bracteoles 2, similar but smaller than bracts. Perianth yellowish, white inside; tube 0.5–1 mm long, with glands lacking or no more than small swellings between the lobes; lobes triangular, 0.6–1.2 mm long, subacute, margins inflexed, membranous, ± erose. Stamens inserted at base of lobes; filaments 0.2–0.5 mm long, almost hidden by anthers, anthers 0.6–0.8 mm long. Style 0.8–1.6 mm long; stigma reaching top of anthers. Fruit pallid, ellipsoid-globose to globose, 3–4 mm long, 2.5–4 mm in diameter, strongly ribbed and reticulate, the reticulations rather more strongly raised than the ribs, dendroid-ruminate. Fig. 1/1–3 (page 4).

KENYA. Northern Frontier Province: Ol Doinyo Lengio, 20 Dec. 1958, *Newbould* 3295!; Machakos District: Chyulu Hills, 13 Dec. 1991, *Luke* 2978!; Kwale District: Shimba Hills National Reserves, Pengo Hill, [1970], *Faden, Evans & Mahasi* 70/821!

TANZANIA. Kilimanjaro, Himo Hill, late 1893 or early 1894, *Volkens* 1712; Mbeya District: Songwe R. gorge by Hot Springs, 27 May 1990, *Carter, Abdallah & Newton* 2490!

DISTR. **K** 1, 4–7; **T** 2, 3 (fide Peter), 7, 8; Congo (Kinshasa), Ethiopia, Angola, Zambia, Malawi, Mozambique and Zimbabwe

HAB. Grassland, sometimes after fires, variously recorded on rocks, limestone, sand and lava ash; 60–2250 m

SYN. *T. andongense* Hiern, Cat. Afr. Pl. Welw. 1: 937 (1900); Baker & Hill in F.T.A. 6(1): 421 (1911). Type: Angola, Pungo Andongo, near Mopopo, *Welwitsch* 6434 (BM!, holo.; K!, iso.)

NOTE. Easily recognised by its winged stems.

6. **Thesium brachyanthum** *Baker* in K.B. 1910: 182 (1910); Baker & Hill in F.T.A. 6(1): 420 (1911). Type: Malawi, Nyika Plateau, Chitipa [Fort Hill], *Whyte* s.n. (K!, holo.)

Annual herb with a small taproot; stem up to 30–60 cm tall, virgately branched above, nearly leafless; ribs raised and terminating in midribs of scale leaves and bracts. Leaves: developed leaves only 2 near base of stem, opposite, linear, ± 15 × 0.4–0.5 mm, acute, involute above; scale leaves closely appressed, lanceolate, ± 1.5–3 × 1–1.5 mm, acuminate, midrib strongly raised beneath. Flowers either solitary or in very compact 3-flowered cymules, sessile, 10–30 in laxly branched spikes that may grow out into sterile scaly tips; bracts lanceolate, 1.5–3.5 × 1.7–2 mm, acuminate, margins membranous, somewhat erose, midrib strongly raised on outer surface; bracteoles 2, similar, sometimes slightly larger. Perianth yellow-green; tube 0.4–0.5 mm long, with glands no more than slight swellings below sinuses of lobes; lobes triangular, 0.8–1.1 mm long, subacute, margins inflexed, sometimes slightly erose near base. Stamens inserted at base of perianth lobes; filaments 0.3–0.5 mm long, nearly hidden by anthers or visible; anthers 0.4–0.7 mm long. Style ± 0.8–1 mm long; stigma reaching top of anthers. Fruit bright red-brown, obovoid, 3–4 mm long, 2.5–3 mm in diameter, strongly ribbed and reticulate. Fig. 1/4, 5 (page 4).

TANZANIA. Ufipa District: 12 km on Namanyere–Chala road, 1 May 1997, *Bidgood et al.* 3622!

DISTR. **T** 4; Zambia, Malawi
HAB. Seasonally inundated *Loudetia* grassland on dark grey sandy loam; 1550 m

NOTE. *Thesium brachyanthum* has the aspect of *T. subaphyllum*, but is easily distinguished by its ribbed but not winged stems, larger and much stouter bracts, and different fruits.

7. **Thesium microphyllum** *Robyns & Lawalrée* in B.J.B.B. 31: 518 (1961). Type: Burundi, Mosso, Kininya, *Michel & Reed* 1525 (BR, holo.)

Perennial herb; rootstock slender, bearing 1–several stems at the apex and later scattered along slender lateral rhizomes; vegetative buds inconspicuous, cone-like, 1–2 mm long, with scales mostly less than 0.5 mm long; stems green, erect, 6–35 cm tall, slender, simple to virgately branched; ribs evident almost to base of the stem, extending upwards into midrib of scales and bracts. Leaves all reduced to scales, grading upwards into bracts, ± scarious, ovate to lanceolate, 1–3 × 0.7–1.2 mm, acute or subacute, slightly erose. Flowers solitary in each bract, sessile, ± 8–20, laxly arranged in spikes starting near base of stem and sometimes extended as sterile scaly shoots above; bracts ovate-lanceolate, 1.5–2.5 × 0.7–1.2 mm, acuminate, keeled, slightly fimbriate or ciliate; bracteoles 2, similar but often a little longer. Perianth yellow-green, white inside; tube 1.5–2.5 mm long, with elongate linear-elliptic glands from near base to near sinuses of lobes, sometimes obscure; lobes linear-triangular, 2–2.8 mm long, with a small hood 0.2–0.3 mm long, margins inflexed, fimbriate near apex inside or all along the margins. Stamens inserted at base of perianth lobes; filaments 1 mm long, visible below the anther; anthers 0.7–1 mm long. Style 3.5–4.5 mm long; stigma reaching base of hoods. Fruit buff to light brown, ovoid-globose, 2.5–3 mm long, 2–2.5 mm in diameter, strongly ribbed and reticulate.

UGANDA. Bunyoro District: Bujenje, Feb. 1943, *Purseglove* 1284!; Masaka District: ± 1.5 km from Katera on Kiebbe road, 1 Oct. 1953, *Drummond & Hemsley* 4508!; Mengo District: Bugerere, Busana, Mar. 1932, *Eggeling* 253!
KENYA. Uasin Gishu, May 1932, *Harvey* 162!; Trans-Nzoia District: 5 km S of Kitale, Kitale Prison Farm, 8 May 1971, *Mabberley & Tweedie* 1121!; Nairobi, between Hillcrest Secondary School and Ngong Road Race Course, 1 Jan. 1980, *M.G. Gilbert* 5813!
DISTR. **U** 2, 4; **K** 3, 4; Burundi
HAB. Short edaphic grassland subject to burning, by lakes, swamps or in depressions (vleis), on sandy or clay soils; 1050–1800 m

SYN. [*T. ussanguense* sensu Agnew, U.K.W.F.: 336 (1974), *non* Engl.]
T. sp. C sensu Agnew, U.K.W.F., ed. 2: 159, t. 58 (1994)

NOTE. Characterised by the slender leafless stems bearing long spikes of small flowers from near the base, the perianth tube relatively long, two-thirds or more the length of the lobes.

8. **Thesium stuhlmannii** *Engl.*, P.O.A. C: 168 (1895); Baker & Hill in F.T.A. 6(1): 422 (1911), excl. specim. *Whyte* 321; F.D.-O.A. 2: 142 (1932); T.T.C.L.: 552 (1949); A.G. Miller in Fl. Eth. & Erit. 3: 380 (1989); U.K.W.F., ed. 2: 159 (1994). Lectotype, chosen here: Tanzania, Bukoba District, Itolio [Itiolo], *Stuhlmann* 934 (B!, lecto.; BM!, K!, isolecto.)

Perennial herb; taproot swollen and sometimes 2–several-fid to support a densely packed bunch of stems without very evident vegetative buds (some small ones are present), occasionally forming slender rhizomes bearing more stems; stems green or orange-green, numerous, erect, 10–60 cm tall, branched from near the base upwards; ribs strongly raised except near base, extending into midrib of scales and bracts. Leaves all reduced to scales, small at base of stem, grading upwards into bracts, mostly lanceolate, lowermost sometimes more oblong-obovate, 1–2.5 × 0.3–1 mm, acute to apiculate, margins scarious and sometimes ± erose, ± keeled. Flowers sessile, 1–2 in each bract, mostly crowded on condensed short shoots, with up to 10 flowers on shoots 0.5–1 cm long, but other flowers in elongate spikes, the bracts more distant below; bracts acuminately ovate or lanceolate, 2–3 × 1.5–2 mm, apiculate, with scarious margins, keeled; bracteoles 2, similar. Perianth yellow-green outside, yellow or whitish within; tube 1–1.2 mm long, with small swellings below sinuses of lobes; lobes narrow, tapering to a hooded tip, 1.8–2.2 mm long, margins slightly inflexed, glabrous; hood 0.1 mm long. Stamens inserted on lower part of perianth lobes; filaments conspicuous, 0.6–0.8 mm long; anthers 0.3–0.5 mm long. Style 2.8–3 mm long, nearly as long as perianth. Fruit buff to pale brown, ovoid-globose to ellipsoid-globose, 1.2–2 mm long, 1–1.5 mm in diameter, ribbed but not strongly reticulate.

UGANDA. Karamoja District: Mt Morongole, July 1965, *J. Wilson* 1654!; Busoga District: Buyende [Buyindi] Hill, 22 May 1951, *G.H.S. Wood* 195!
KENYA. Turkana District: near summit of Kachagalau Mt, 22 Apr. 1959, *Osmaston* 4461!; Machakos District: Chyulu Hills, near Half Crater, 22 Oct. 1990, *Luke* 2458!; N Kavirondo District: Suna, Sept. 1933, *Napier* 3092 in *C.M.* 5332!
TANZANIA. Bukoba District: Kakindu, Oct. 1931, *Haarer* 2279!
DISTR. **U** 1, 3; **K** 2–6; **T** 1; Ethiopia
HAB. Short grassland subject to fire, sometimes in rocky places or on lava ash; 1050–2750 m

NOTE. Differs from *T. fimbriatum* by the glabrous perianth lobes and small glands not forming teeth between the lobes. It occurs in drier burned grassland.
Baker & Hill in F.T.A. (1911) attributed part of the collection of *Whyte* 321 from Malawi to *T. stuhlmannii*, but Hilliard in F.Z. (ined.) shows that all the Malawi material belongs to *T. fimbriatum*.

9. **Thesium fimbriatum** *A.W.Hill* in K.B. 1910: 184 (1910), *non* A.W.Hill in K.B. 1915: 27 (1915); Baker & Hill in F.T.A. 6(1): 422 (1911). Type: Malawi, between Khondowe and Karonga, *Whyte* 321 in part (K!, holo.)

Perennial herb, at first appearing annual; taproot thickening a little at crown, tending to branch and produce slender rhizomes; stems pale to dull green, sometimes dull orange-bronze tinged, 1–several, 25–85 cm tall, erect, simple to virgately branched, the branches sharply ascending, the flowering branches later extended into scaly tips; ribs raised except near base of stem, terminating in midribs of scale leaves and bracts. Leaves all reduced to scales, closely appressed, lanceolate, 0.5–2 × 0.4–1.6 mm, increasing in size upwards, distant below, becoming more crowded upwards and passing into bracts, acuminate, minutely erose-ciliate, midrib strongly raised on outer surface. Flowers sessile, with a bract and 2 bracteoles or with an involucre of 3–5 bracteoles, up to 10 crowded on a condensed short shoot, these either simple or branched, or sometimes in long many-flowered spikes, the flowers ± crowded below, distant above, both condensed and long spikes sometimes present on one plant; lower bracts ovate-acuminate, 2–3.5 × 1–2 mm, margins membranous, minutely erose-ciliate, keeled by strongly raised midrib; bracteoles similar to bracts. Perianth greenish yellow, whitish inside; tube 0.5–1 mm long, with 5 tiny (± 0.1 mm) tooth-like glands alternating with the lobes; lobes (4–)5, subspathulate, 1.5–2.7 mm long, apex shortly hooded, subobtuse to acute, margins inflexed, fringed with long white hairs; hood 0.1–0.25(–0.4) mm long. Stamens inserted near base of perianth lobes; filaments conspicuous, 0.9–1.5 mm long, lower part adnate to perianth lobe; anthers 0.3–0.5 mm long. Style ± 1.2–2.6 mm long; stigma reaching ± to middle or top of anthers. Fruit buff-coloured to light reddish brown, ovoid-globose, 1.6–2 mm long, 1.2–2 mm in diameter, strongly ribbed, the 5 intermediate ribs often forking once near base, reticulations few and often obscure. Fig. 1/8–9 (page 4).

TANZANIA. Ufipa District: 13 km on Sumbawanga–Tunduma road, 20 Feb. 1994, *Bidgood, Mbago & Vollesen* 2356!; Iringa District: Mufindi, Ngowasi Lake, 22 Mar. 1962, *Polhill & Paulo* 1830!; Tunduru District: ± 1.5 km E of R. Mawese near Puchapucha, 19 Dec. 1955, *Milne-Redhead & Taylor* 7819!

DISTR. **T** 4, 7, 8; Congo (Katanga), Zambia, Malawi, Mozambique
HAB. Wet grassland and swampy areas, less often in wooded grassland on sandy loams; 250–1900 m

SYN. *T. shabense* Lawalrée in B.J.B.B. 55: 23 (1985). Type: Congo (Kinshasa), Katanga, Kundelungu Plateau, near W source of R. Lutshipuka, *Lisowski, Malaisse & Symoens* 4276 (BR, holo.; POZG, iso.)

NOTE. Closely related to *T. stuhlmannii*, but occurs further south and generally in wetter habitats. *T. schliebenii* is sympatric and differs as indicated below.

10. **Thesium schliebenii** *Pilg.* in N.B.G.B. 11: 397 (1932). Type: Tanzania, Njombe District, Lupembe, *Schlieben* 757 (B, holo., seen by Hilliard in 1989; BR, iso.)

Perennial herb, with narrow woody rhizomes up to 5 cm long; stems several from each rhizome, the whole forming a many-stemmed tuft; stems erect, 25–60 cm tall, simple to virgately branched, flowering branches sharply ascending, later extending into sterile scaly tips; ribs strongly raised and extending into midribs of scale leaves and bracts. Leaves all reduced to scales, closely appressed, lanceolate, 1–2.5 × 0.4–1.2 mm, increasing in size upwards and passing into floral bracts, acuminate, midrib strongly raised on outer surface. Flowers sessile, with a bract and 2 similar bracteoles or an involucre with 3–5 bracts, usually up to 6 flowers crowded on very condensed short shoots, but sometimes also spaced on elongate shoots, both types occurring on the same plant; bracts and bracteoles ovate-acuminate, 2–3 × 1–1.4 mm, margins membranous, minutely erose-ciliate, keeled by strongly raised midrib. Perianth greenish white or yellow; tube 1 mm long, 5 tiny (up to 0.1 mm) tooth-like glands alternating with the lobes; lobes linear, 3–3.6 × 0.4–0.5 mm, tapered to a long hood, margins inflexed, erose in upper part, densely white-bearded in lower part; hood 0.6–0.8 mm long. Stamens inserted near base of perianth lobes; filaments conspicuous, 1.5–2.2 mm long, shortly adnate to lobe; anthers 0.4–0.7 mm long. Style 2.8–3.5 mm long; stigma reaching base of hood. Fruit buff-coloured to pale brown, globose, 2 mm in diameter, very strongly ribbed, very weakly reticulate.

TANZANIA. Iringa District: Kigogo Forest, escarpment below Fox's viewpoint, 27 Dec. 1986, *Lovett & Congdon* 1202!; Njombe District: Njombe, 12 Dec. 1931, *Lynes* D.d.138! & Mwakete–Njombe road, 17 Jan. 1957, *Richards* 7878!
DISTR. **T** 7; Congo (Katanga), Zambia
HAB. Wet grassland or swamps; 1600–2100 m

SYN. *T. schaijesii* Lawalrée in B.J.B.B. 55: 22 (1985). Type: Congo (Kinshasa), Katanga, 92 km S of Kolwezi, source of R. Musinga, *Schaijes* 1588 (BR, holo.)

NOTE. *T. schliebenii* is closely related to *T. fimbriatum*, differing principally in the longer hood of the perianth lobes.

11. **Thesium fastigiatum** *A.W.Hill* in K.B. 1910: 183 (1910); Baker & Hill in F.T.A. 6(1): 427 (1911). Type: Mozambique, on a ridge at the back of Mendwe Chujala-Yala, *Johnston* 316 (K!, holo.)

Perennial herb; rootstock woody, irregularly shaped, 1–4 cm across, with many very short closely scaly vegetative buds that eventually elongate into stems; buds 2–4 mm long, the scales straw-coloured, subulate, 1–2 mm long; stems pale brown at base, green above, numerous, erect or decumbent, 30–60 cm high, fastigiately well-branched from the base or a little higher, branches irregularly racemose-cymose, ascending at angle of ± 45°, eventually almost filiform, up to 20–30 cm long, minutely scabridulous to glabrous, very shallowly ribbed by decurrent scale leaves. Leaves usually all reduced to scales, those at extreme base of stem ovate or ovate-lanceolate,

1.5–2.5 mm long, those subtending primary branches lanceolate or ovate-lanceolate, up to 2.5–7 × 0.6–1.6 mm, occasionally with developed linear leaves interspersed and up to 25 × 0.8 mm, cuspidate, thick-textured, margins smooth to minutely ciliolate. Flowers sessile, nearly always on a scaly shoot (primary branch from leaf axil) initially branching racemosely rather than cymosely, irregularly cymose thereafter, and occasionally (always mixed with scaly shoots) a nude peduncle terminating in a flower may develop from the leaf axil, and is soon overtopped by two scaly cymose branches, many branches eventually running up into long filiform scaly apices; bracts ± 5 forming an involucre below the flower, ovate-acuminate, 0.8–1.5 × 0.5–0.9 mm, margins ciliolate. Perianth green with orange-buff tips, white inside; tube 0.5–1 mm long, external ellipsoid glands very obscure; lobes oblong-lanceolate, 0.8–1.7 mm long, subacute, hood 0.3 mm long, margins inflexed, ± erose. Stamens inserted at base of perianth lobes; filaments 0.3–0.5 mm long, ± hidden by anthers; anthers 0.3–0.7 mm long. Style 1–1.6 mm long; stigma reaching almost to apex of hood. Fruit buff to pale brown, ellipsoid-globose, 2.5–3 mm long, 2–2.5 mm in diameter, strongly ribbed and reticulate. Fig. 1/10–12 (page 4).

TANZANIA. Ufipa District: between Kasanga and Sumbawanga, 30 Mar. 959, *McCallum Webster* T.501!; Songea District: 16 km W of Songea, 2 Jan. 1956, *Milne-Redhead & Taylor* 8110!; Masasi District: Ndanda Mission, 7 Mar. 1991, *Bidgood, Abdallah & Vollesen* 1827!

DISTR. **T** 4, 8; Congo (Kinshasa), Zambia, Malawi, Mozambique, Zimbabwe

HAB. *Brachystegia* woodland and upland grassland, generally on eroded or bare ground on slopes, on loam, sand or gravel; 450–2200 m

SYN. *T. tamariscinum* A.W.Hill in K.B. 1910: 188 (1910); Baker & Hill in F.T.A. 6(1): 431 (1911). Lectotype, chosen by Hilliard in F.Z.: Malawi, Khondowe to Karonga, 2000–6000 ft., July 1896, *Whyte* s.n. (K!, lecto.)

T. luembense Robyns & Lawalrée in B.J.B.B. 17: 146 (1944) & in F.C.B. 1: 302 (1948). Type: Congo (Kinshasa), Katanga, valley of Little Luembe, *Hock* s.n. (BR, holo.)

NOTE. All the Tanzanian material seen has scale leaves only, but some material from further south has linear leaves interspersed, e.g. *Nuvunga* 569, from the Niassa Province of Mozambique at Lichinga, has developed linear leaves to 10 mm long subtending some lateral branches, but the general appearance of the plant is still scaly.

See note under *T. unyikense*.

12. **Thesium unyikense** *Engl.* in E.J. 30: 306 (1901); Baker & Hill in F.T.A. 6(1): 428 (1911); U.K.W.F., ed. 2: 159, t. 58 (1994). Type: Tanzania, Mbeya Distr., Unyika, Fingano village, *Goetze* 1377 (B, holo., seen by Hilliard in 1989; BM!, BR, EA, iso.; K!, frag.)

Perennial herb; rootstock woody, thickly disk-shaped, 2–4 cm across, crowned with very many short closely scaly vegetative buds that elongate into stems; buds 3–5 mm long, with brown triangular scales 1–2 mm long; stems grey-green, numerous, erect or decumbent, 5–40 cm tall, soon fastigiately branched, branches irregularly cymose, ascending at angle of ± 45°, seldom more than 5 cm long; ribs low and broad, decurrent from scale leaves, often minutely pustulate. Leaves all reduced to scales, those at extreme base of stem triangular-lanceolate, 1.5–3 mm long, those subtending branches lanceolate to ovate-lanceolate, 1.5–6 × 0.8–1.6 mm, cuspidate, thick-textured, margins membranous, ciliolate. Flowers sessile, initially solitary at tips of nude peduncles ± 4–30 mm long (measured from axil of bract to first flower), often 2 peduncles from 1 axil, later overtopped by 1 or occasionally 2 branches arising immediately or shortly below flower, branch either nude and terminating in a flower, or bracteate, terminating in a flower and with a floral bud in the axil of each bract; this branching may be repeated again; bracts ± 5, forming an involucre below the flower, lanceolate to ovate-lanceolate, 1–2 × 0.5–1 mm, acuminate to cuspidate, margins ciliolate. Perianth yellow-green outside, white to cream inside; tube (0.6–)0.8–1.2 mm long, external glands ellipsoid, whitish, sometimes not very evident; lobes oblong-lanceolate, 1–1.8 mm long, obtuse, margins inflexed, ± erose; hood 0.25–0.3 mm long. Stamens inserted at base of perianth lobes; filaments 0.25–0.5 mm long, hidden by anthers; anthers 0.5–0.8 mm long. Style 1–2 mm long; stigma reaching nearly to top of perianth lobes. Fruit buff to pale brown, ellipsoid to ovoid-globose, 2.5–3 mm long, 2.5–2.75 mm in diameter, strongly ribbed and reticulate. Fig. 1/13–15 (page 4).

KENYA. Trans-Nzoia District: Kitale, Mar. 1966, *Tweedie* 3255! & Kitale, Milimani, Mar. 1968, *Tweedie* 3527!
TANZANIA. Ufipa District: Chapota, 4 Dec. 1949, *Bullock* 2012!; Mbeya District: Mbozi [Mbosi], Nov. 1928, *Haarer* 1622!; Njombe District: 29 km from Itulo sheep farm on road to Njombe, 13 Nov. 1966, *Gillett* 17817!
DISTR. **K** 3; **T** 4, 7, 8; Zambia, Malawi, Mozambique, Zimbabwe, ? Angola
HAB. Upland grassland and damp places in *Brachystegia* woodland or by streams; 1500–2300 m

NOTE. The flowers of the two gatherings from Elgon are at the upper end of the range of flower size, but are matched by some specimens from the Flora Zambesiaca area.
T. unyikense and *T. fastigiatum* (together with the more hairy *T. thamnus*) belong to a very characteristic group of species with the flowers subtended by an involucre of bracts. In *T. fastigiatum* the flowers terminate a scaly shoot, with in addition sometimes a few additional flowers on short axillary naked peduncles. In *T. unyikense* the flowers characteristically terminate a naked peduncle, often with one to few scales below the cluster of bracts. Buds grow out from the axils of a lower bract and/or the scale leaf below, forming a new subnaked peduncle ascending next to the previous one, the pattern then repeated. In both species fertile shoots can grow out into long sterile scaly tips. These distinctions are usually evident and define the species satisfactorily in the Flora Zambesiaca area, but in southern Tanzania the pattern becomes a little blurred on the Ufipa.

13. **Thesium thamnus** *Robyns & Lawalrée* in B.J.B.B. 31: 522 (1961). Type: Congo (Kinshasa), Katanga, 40 km SE of Lubumbashi, Mushoshi road, *Schmitz* 3278 (BR, holo.)

Perennial herb; rootstock woody, irregularly shaped, 1–3 cm across, with very many short scaly buds that eventually elongate into stems; buds 2–5 mm long, the scales straw-coloured (the outer ones sometimes brown-tinged), subulate, ciliate; stems grey-green, numerous, ascending to erect, 10–30 cm high, well-branched and floriferous almost from the base, the branches ascending at an angle of ± 45°, relatively short; stems hairy at least on the lower part, weakly ribbed by decurrent scale leaves; hairs stiff, spreading, parallel-sided, appearing almost glandular, 0.1–0.5 mm long, shorter upwards but extending on to back of bracts. Leaves all reduced to scales, lanceolate-acuminate, 2–7 × 1–2 mm, ciliate and hairy outside. Flowers initially solitary at tips of nude peduncles, the peduncles 2–20 mm long and often two from one axil, later overtopped by 1–2 branches arising immediately or shortly below the terminal flower; these may branch again, producing a nude peduncle terminating in a flower or a sterile scaly shoot; bracts ± 4–5, forming an involucre below the flower, ovate-lanceolate, acuminate, 1.3–2.5 × 0.7–1.1 mm, ciliate and hairy on the back. Perianth white inside; tube 0.5–1 mm long, external glands obscure, ellipsoid; lobes 5(–6), oblong, 1.1–1.4 mm long, obtuse with a hood ± 0.3 mm long, margins inflexed, erose. Stamens inserted at base of the perianth lobes; filaments 0.1–0.5 mm long, largely hidden by anthers; anthers 0.3–0.8 mm long. Style 1–1.8 mm long; stigma reaching to top of the anthers or apex of the hood. Fruit buff to pale brown, 2.8–3.5 mm long, 2–2.5 mm in diameter, strongly ribbed and reticulate.

TANZANIA. Iringa District: Iringa, 26 Dec. 1964, *Tallantire* T. 908! & Mafinga [Sao Hill], 29 Sept. 1968, *B.J. & S. Harris* 2389!
DISTR. **T** 7; Congo (Katanga), Zambia and Zimbabwe
HAB. Grassland subject to burning; 1500–1900 m

SYN. *T. unyikense* Engl. var. *puberulum* R.E.Fries, Wiss. Ergebn. Schwed. Rhod.-Kongo-Exped.: 21 (1914). Type: Zambia, Bwana Mkubwa, *R.E. Fries* 462 (UPS, holo.)

NOTE. Readily distinguished from *T. unyikense* and *T. fastigiatum* by the hairs, which are conspicuous on the bracts, scale leaves and at least the lower part of the stem.

14. **Thesium radicans** *A.Rich.*, Tent. Fl. Abyss. 2: 235 (1850); Engl., Hochgebirgsfl.: 250 (1892); Baker & Hill in F.T.A. 6(1): 415 (1911); A.G. Miller in Fl. Eth. & Erit. 3: 380, fig. 116.1.5–7 (1989); U.K.W.F., ed. 2: 159 (1994). Types: Ethiopia, near Adua, Dagabur church, *Schimper* I.168 (P, syn.; B!, BM!, BR, K!, isosyn.) & Endchedcab, *Schimper* II.1130 (P, syn.; B!, BR, isosyn.) & Ouodgerate, *Petit* s.n. (P, syn.)

Perennial herb forming dense mats; rhizomes slender, ± 1 mm in diameter; stems slender, erect, 3–10 cm tall, with leafy branches that creep up to 30 cm, bearing further short erect flowering branches; ribs not well formed, partly decurrent from base of leaves and partly from the midribs of leaves and bracts. Leaves almost all well developed, with only a few small scale leaves near the base of the stems, green, linear, 10–20 × 0.7–1.2 mm, acute, mucronate, slightly tapered basally, with a cartilaginous margin and the midrib evident beneath, sometimes slightly scabridulous. Flowers 1–6, single in each bract or 2–3 in a cymule, terminal or leaf-opposed on the short stems or on lateral branches 3–15 cm long, sometimes with axillary buds growing into a leafy extension; bracts shortly or wholly adnate to the pedicel or peduncle, linear, leafy, 3–8 × 0.8–1.2 mm; bracteoles 2(–3), a little smaller; pedicel 1–2 mm long. Perianth greenish outside, white inside; tube 0.5–0.6 mm long, the glands slight elliptic swellings below the sinuses of the lobes or obscure; lobes triangular, 0.6–0.8 mm long, subacute, slightly hooded, bearded at apex inside. Stamens inserted at base of perianth lobes; filaments 0.05–0.1 mm long; anthers 0.2–0.3 mm long. Style 0.6–0.7 mm long; stigma reaching to level of anthers. Fruit orange to bright red when ripe, globose-ellipsoid to subglobose, 2–3 mm across, thinly fleshy, with faint venation when dried; stipe 0.5–0.8 mm long.

KENYA. Naivasha District: SW of Lake Naivasha, shore near Y.M.C.A. Camp, 19 Apr. 1968, *Mwangangi* 758!; Kiambu District: John Steven's Farm, Mar. 1957, *Bally* 11411!; Masai District: Olokurto Mau area, 1 June 1961, *Glover, Gwynne & Samuel* 1515!

TANZANIA. Arusha District: Meru Crater, 28 May 1968, *Renvoize & Abdallah* 2411!; Moshi District: Useri, Jan. 1929, *Haarer* 1792! & Machame, Central Girls' School, 22 Feb. 1955, *P. Huxley* 125!

DISTR. **K** 1, 3, 4, 6; **T** 2; Ethiopia, Eritrea, Yemen and Saudi Arabia (record for Somalia based on a misidentification)

HAB. Short grassland, grazed places, rock crevices; 1350–2650 m

SYN. *T. wightianum* Wight var. *radicans* (A.Rich.) DC., Prodr. 14: 647 (1857)
T. prostratum Peter, F.D.-O.A. 2: 142, Anh. 12, t. 16/2 (1932). Type: Tanzania, Masai District, Embulbul [Bulbul], to Munge [Lemunge], *Peter* 43038 (B†, holo.)
T. sp. B sensu Agnew, U.K.W.F.: 238 (1974)

NOTE. Easily recognised by the bright red fruits borne near the ground, the plant forming a mat in short grassland.

15. **Thesium schweinfurthii** *Engl.*, P.O.A. C: 168 (1895); Baker & Hill in F.T.A. 6(1): 416 (1911); A.G. Miller in Fl. Eth. & Erit. 3: 380, fig. 116.1.11–13 (1989); U.K.W.F., ed 2: 159 (1994), pro minore parte. Type: Sudan, Jur [Djur], Molmul R., *Schweinfurth* 1909 (B, holo., seen by Hilliard in 1989; K!, iso.)

Perennial herb; rootstock generally narrow and elongate, sometimes (particularly in the Rift Valley region of Kenya) 1–several-forked at the apex and ± shortly rhizomatous; vegetative buds, if evident, 1–2 mm long; stems many, tufted from the crown (top of the rootstock), sprawling with age, 10–40 cm long, well-branched from the base, leafy, weakly ribbed by decurrent leaves. Leaves almost all well developed, only a few small scale leaves at the base, green, ascending, linear, 10–20(–25) × 0.8–1.2(–2) mm, somewhat shorter at the base, apex apiculate, margins pale, cartilaginous, smooth, with midrib evident. Flowers solitary in the axils of the bracts, 1–5(–7) laxly arranged at the tips of all the branchlets, oldest at apex; bracts adnate to the entire pedicel, leaf-like, linear, ± 12–25 × 0.8–1.2 mm; bracteoles 2, ± 3.5–11 × 0.5–0.8 mm; pedicels 0.5–5 mm long. Perianth green outside, white inside; tube 0.5–0.7 mm long, external glands ellipsoid below sinuses of lobes or inconspicuous; lobes triangular, 0.8–1 mm long, acute, white-bearded from apex. Stamens inserted at base of perianth lobes; filaments ± 0.25 mm long, visible; anthers 0.25–0.3 mm long. Style 0.2–0.7 mm long; stigma reaching about halfway up anthers or a little above. Fruit pale yellowish or reddish brown, subglobose, 2.2–2.5(–3) mm long, 2–2.5(–2.8) mm in diameter, strongly ribbed and reticulate; stipe 1–1.5 mm long.

UGANDA. Acholi/Karamoja District: Imatong Mts, Langia, Apr. 1943, *Purseglove* 1384!; Ankole District: Kamukaki–Rugongo, 16 Dec. 1993, *Rwaburindore* 3639!; Teso District: Serere, Nov.–Dec. 1931, *Chandler* 86!

KENYA. Northern Frontier Province: Mt Kulal, 9 Oct. 1947, *Bally* 5570!; Trans-Nzoia District: NE Elgon, Apr. 1957, *Tweedie* 1447!; Naivasha District: ± 32 km from Kikuyu on direct road to Narok down Rift via Wanyaga, 20 Jan. 1963, *Verdcourt* 3555!

TANZANIA. Ngara District: Bugufi, Nyamyaga [Nyamiaga], 23 Aug. 1960, *Tanner* 5105B!; Dodoma District: Kazikazi, 21 Feb. 1934, *B.D. Burtt* 5066!; Rungwe District: Ipyana, 30 Dec. 1912, *Stolz* 1752!

DISTR. **U** 1–3; **K** 1–3, 5, 6; **T** 1, 4, 5, 7; Nigeria, Congo (Kinshasa), Rwanda, Burundi, Sudan, Ethiopia, Zambia, Malawi

HAB. Upland grassland, wooded and seasonally flooded grassland, *Brachystegia* woodland and mixed bushland, sometimes after fires or in disturbed places, on loam or clay; 1050–2300 m

SYN. *T. schweinfurthii* Engl. var. *djurense* Engl., P.O.A. C: 168 (1895), *nom. superfl.* Type as for species

T. sphaerocarpum Robyns & Lawalrée in B.J.B.B. 31: 513 (1961). Type: Rwanda, Rubona, *Michel* 4655 (BR, holo.; K!, iso.)

T. schmitzii Robyns & Lawalrée in B.J.B.B. 31: 514, fig. 57 (1961). Type: Congo (Kinshasa), Katanga, 10 km SSE of Lubumbashi [Elisabethville], *Schmitz* 3356 (BR, holo.)

T. sp. A sensu U.K.W.F.: 337, fig. on p. 337 (1974), pro parte, & ed. 2: 159, t. 58 (1992)

NOTE. *T. schweinfurthii* occurs essentially west of the Great Rift Valley and *T. mukense* to the east. Material from the Rift Valley floor from Naivasha to Ngorongoro is mostly best attributed to *T. schweinfurthii*, though these populations tend to have small bracts, flowers up to 7 per raceme at the upper end of the size range for the species, and the crown of the rootstock tends to branch and extend as short thick rhizomes. Both species occur around the Ngong Hills.

16. **Thesium mukense** *A.W.Hill* in K.B. 1910: 186 (1910), excl. specim. *Buchanan* 45; Baker & Hill in F.T.A. 6(1): 417 (1911), excl. specim. *Buchanan* 45. Lectotype, chosen by Hilliard in F.Z.: Kenya, Machakos District, Mukaa [Muka], *Kassner* 943 (K!, lecto.; BM!, isolecto.)

Perennial herb with woody rhizomes ± 3–4 mm in diameter; vegetative buds ± 2 mm long; stems yellow-green to grey-green or somewhat glaucous, in small tufts, 10–30 cm tall, erect, well-branched from the base or a little higher, leafy, shallowly ribbed by decurrent leaves. Leaves rather fleshy, ± ascending, primary ones on main stems linear, often rather curved, 10–45 × 0.6–1.2 mm, shorter upwards and passing into bracts, apex apiculate, smooth. Flowers solitary in the axil of each bract, 5–14 racemosely arranged at the tips of the branchlets, the inflorescence often terminating in a 3-flowered cymule; bracts adnate to entire pedicel, linear, leaf-like, 5–17 × 0.6–0.8 mm; bracteoles 2, ± 1.5–5 × 0.4–0.6 mm; pedicels 1–3.5(–5) mm long. Perianth green or orange-buff outside, white inside; tube 0.6–1 mm long, external glands ellipsoid, inconspicuous; lobes triangular, 1.2–1.7 mm long, acute, white-bearded from apex. Stamens inserted at base of perianth lobes; filaments 0.3–0.5 mm long, hidden by anthers; anthers 0.5–0.8 mm long. Style 0.5–0.7 mm long; stigma reaching top of tube or a little above. Fruit light yellowish brown to reddish brown, ellipsoid-globose to globose, 3–4 mm long, 2.5–3.5 mm in diameter, strongly ribbed and reticulate; stipe 0.5–1 mm.

KENYA. N Nyeri District: 7 km on Nyeri–Kiganjo road, 28 May 1974, *Faden & Evans* 74/622!; Nairobi District: Langata, between Hillcrest Secondary School and Ngong Road Forest, 24 May 1980, *M.G. Gilbert* 5953!; Teita District: near Bura, 3 Dec. 1998, *Luke et al.* 5541!

TANZANIA. Masai District: Malanja, 14 Dec. 1962, *Newbould* 6367!; Moshi District: Engare [Ngari] Nairobi, 4 July 1943, *Greenway* 6703!; Handeni District: Gendagenda Forest, 11 Aug. 1991, *Frontier Tanzania* 2413!

DISTR. **K** 4, 6, 7; **T** 2, 3; Mozambique and South Africa (KwaZulu-Natal)

HAB. Upland grassland, wooded grassland and seasonally flooded grassland, areas subject to burning or erosion, descending to lower altitudes along rivers, on loam, clay or lava ash; (300–)1100–2500 m

SYN. *T. sp. A* sensu U.K.W.F.: 337 (1974), pro minore parte quoad "NAN" & ed. 2: 159 (1992), pro parte

[*T. schweinfurthii* sensu Agnew, U.K.W.F., ed. 2: 159 (1964), pro majore parte, *non* Engl.]

NOTE. See note under *T. schweinfurthii*.

17. **Thesium goetzeanum** *Engl.* in E.J. 30: 306 (1904); Baker & Hill in F.T.A. 6(1): 418 (1911); Hill in F.C. 5(2): 181 (1915); F.D.-O.A. 2: 142 (1932); N.E. Br. in Fl. Pl. Ferns Trans.: 459 (1932); T.T.C.L.: 551 (1949). Type: Tanzania, Mbeya District, Unyika, Fingano, *Goetze* 1379 (B!, holo.; BM!, iso.; K!, frag.)

Perennial herb with thick woody rootstock, sometimes with rhizomes; vegetative buds not very evident, ± 2 mm long, with triangular-acuminate scales 1–1.5 mm long; stems tufted from crown, 10–120 cm tall, sparingly to well branched, leafy; ribs strongly raised and extended into midribs of leaves and bracts. Leaves mostly well-developed, with a few scales near base of stems, glaucous, thick-textured, ascending to somewhat spreading and then slightly recurved, linear-lanceolate, 6–20 × 1–1.5mm midway on stem, smaller upwards and passing into bracts, often shorter and slightly broader at base, apex apiculate, base broad, margins pale, cartilaginous, very minutely scabridulous or smooth, midrib raised on both surfaces. Flowers 7–13 racemosely arranged at tips of branches, laxly panicled, usually solitary in each bract axil, occasionally paired or in 3-flowered cymules, sometimes all solitary except for 3-flowered terminal cymule; bracts usually adnate to entire length of pedicel, rarely only to lower half of peduncle, lanceolate, leaf-like, 4–12(–20) × 1–2 mm; bracteoles 2, 2–7.5 × 0.4–1.3 mm; pedicels up to 0.5–2(–4) mm long; peduncles of cymules up to 10 mm. Perianth greenish; tube 0.7–1.2 mm long, external glands ± conspicuous, elliptic, equalling tube; lobes triangular, 1.2–2 × 0.6–0.9 mm, subacute, densely white-bearded from apex. Stamens inserted at base of perianth lobes; filaments 0.2–0.5 mm long, almost hidden by anthers; anthers 0.4–0.8 mm long. Style 0.6–1.1 mm long; stigma reaching nearly to top of anthers. Fruit yellowish buff or slightly reddish buff, ellipsoid-globose, 3–3.5 mm long, 2.5–3 mm in diameter, ribs strongly raised, reticulations rather weakly raised in comparison, becoming stronger only at final maturation. Fig. 1/16 (page 4).

KENYA. N Nyeri District: 28 km E of Timau on road to Meru, 16 Apr. 1979, *Gillett* 21816!

TANZANIA. Mbulu District: Fungos, Oct. 1925, *Haarer* 124B!; Pangani District: Mwera, Langoni, 30 Oct. 1956, *Tanner* 3206!; Buha District: Murungu–Nyamgalika, 20 Aug. 1950, *Bullock* 3192!

DISTR. **K** 4; **T** 2–7; Rwanda, Zambia, Malawi, Mozambique, Zimbabwe and South Africa (Mpumalanga, KwaZulu-Natal)

HAB. Grassland, often in areas subject to fires; 250–2050 m

SYN. *T. schweinfurthii* Engl. var. *laxum* Engl., P.O.A. C: 168 (1895); F.D.-O.A. 2: 141 (1932). Type: Tanzania, Kilimanjaro, below Marangu, *Volkens* 2100 (B†, holo.)
[*T. cymosum* sensu Vollesen in Opera Bot. 59: 63 (1980), *non* A.W.Hill]
[*T. dissitum* sensu Vollesen in Opera Bot. 59: 63 (1980), *non* N.E.Br.]

NOTE. Closely related to *T. schweinfurthii* and *T. mukense*, but the more cymose structure of the inflorescences is generally evident if the plant is considered as a whole.

2. **OSYRIDICARPOS**

A.DC. in DC., Prodr. 14: 635 (1857); Stauffer in Viert. Nat. Ges. Zürich 106: 400, fig. 5A–F (1961)

Scandent shrub; branchlets terete, ribbed. Leaves alternate, 3-nerved, petiolate. Flowers hermaphrodite, in long leafy terminal racemes or panicles, often solitary in the axil of each leaf-like bract, rarely in 3(–7)-flowered dichasia; bract free from peduncle; bracteoles 2, minute. Perianth tube cylindric; lobes 5, valvate. Stamens 5, inserted at base of perianth lobes; anthers attached to face of lobe by a conspicuous tuft of hairs. Disk lobed, distinct. Ovary with 2–3 ovules from apex of erect twisted free-central placenta, the tip of which is drawn out into a point; stigma ± capitate, very obscurely 2–5-lobed. Fruit indehiscent, globose, endocarp bony, obscurely 5-ribbed, outer coat fleshy, crowned with persistent perianth.

A monotypic genus confined to the eastern side of Africa.

Osyridicarpos schimperianus (*A.Rich.*) *A.DC.* in DC., Prodr. 14: 635 (1857); Oliv. in Hooker, Ic. Pl. 15, t. 1443 (1883); Engler, Hochgebirgsfl. Trop. Afr.: 199 (1892) & P.O.A. C: 168 (1895); Baker & Hill in F.T.A. 6(1): 432 (1911); F.D.-O.A. 2: 143 (1932); T.T.C.L.: 550 (1949) & in Mem. N.Y. Bot. Gard. 9: 65 (1954); Stauffer in Viert. Nat. Ges. Zürich 106: 400, fig. 5 (1961); A.G. Miller in Fl. Eth. & Erit. 3: 382, fig. 116.2 (1989); U.K.W.F., ed. 2: 160 (1994); White, Dowsett-Lemaire & Chapman, Evergreen For. Fl. Malawi: 520 (2001). Types: Ethiopia, Adua, Mt Sholoda, *Schimper* I. 404 (P, syn.; B!, BM!, BR, K!, isosyn.) & Tigray, *Quartin Dillon* s.n. (P, syn.; MO, isosyn.)

Shrub scrambling to 1–5 m, pendent or straggling if unsupported, with many ascending branches; branches strongly ribbed, glabrous to coarsely hairy. Leaves dark green, thick-textured, narrowly to broadly elliptic, 6–50 × 2.5–15 mm, smaller upwards, tapering at both ends, apex acute; petiole 1–4 mm long, articulated to a small cushion, generally glabrous or soon glabrescent in East Africa. Flowering shoots 3–30 cm long, mostly with 10–30 flowers; bract leaf-like at base to small and linear at tip; bracteoles 2, opposite, linear-lanceolate, ± 1 mm long. Perianth white to pale yellow or greenish, leathery, with 5 ± distinct ribs running from the centre of each lobe to the ovary, external glands at base of perianth either well developed or not; tube cylindric, 3.5–6.5 mm long; lobes spreading, ovate, 1.2–2.3 mm long, acute, tips slightly hooded. Stamens: filaments 0.8–1.5 mm long; anthers 0.7–1 mm long. Ovary ± turbinate, strongly 5-ribbed; style 4.5–7.7 mm long, cylindric, thick; stigma visible in mouth. Fruit globose, 5–6 mm in diameter, bony but the epicarp fleshy, eventually creamy-white. Fig. 4 (page 22).

UGANDA. Karamoja District: Mt Moroto, May 1963, *J. Wilson* 1396! & foothills of Mt Moroto, June 1963, *Tweedie* 2647! & Lodoketemit catchment area, 13 July 1958, *Kerfoot* 307!
KENYA. Northern Frontier Province: Moyale, 3 July 1952, *Gillett* 13490!; Kiambu District: Muguga North, 21 May 1961, *Verdcourt* 3136!; Teita District: Taita Hills, E of Mwandongo Forest, 6 Nov. 1998, *Mwachala et al.* in *Earthwatch* 816!
TANZANIA. Arusha National Park, Engare Nanyuki R., 11 Apr. 1968, *Greenway & Kanuri* 13451!; Ufipa District: Mbisi Forest, 11 Aug. 1960, *Richards* 13061!; Njombe District: 13 km S of Njombe, 8 July 1956, *Milne-Redhead & Taylor* 11016!

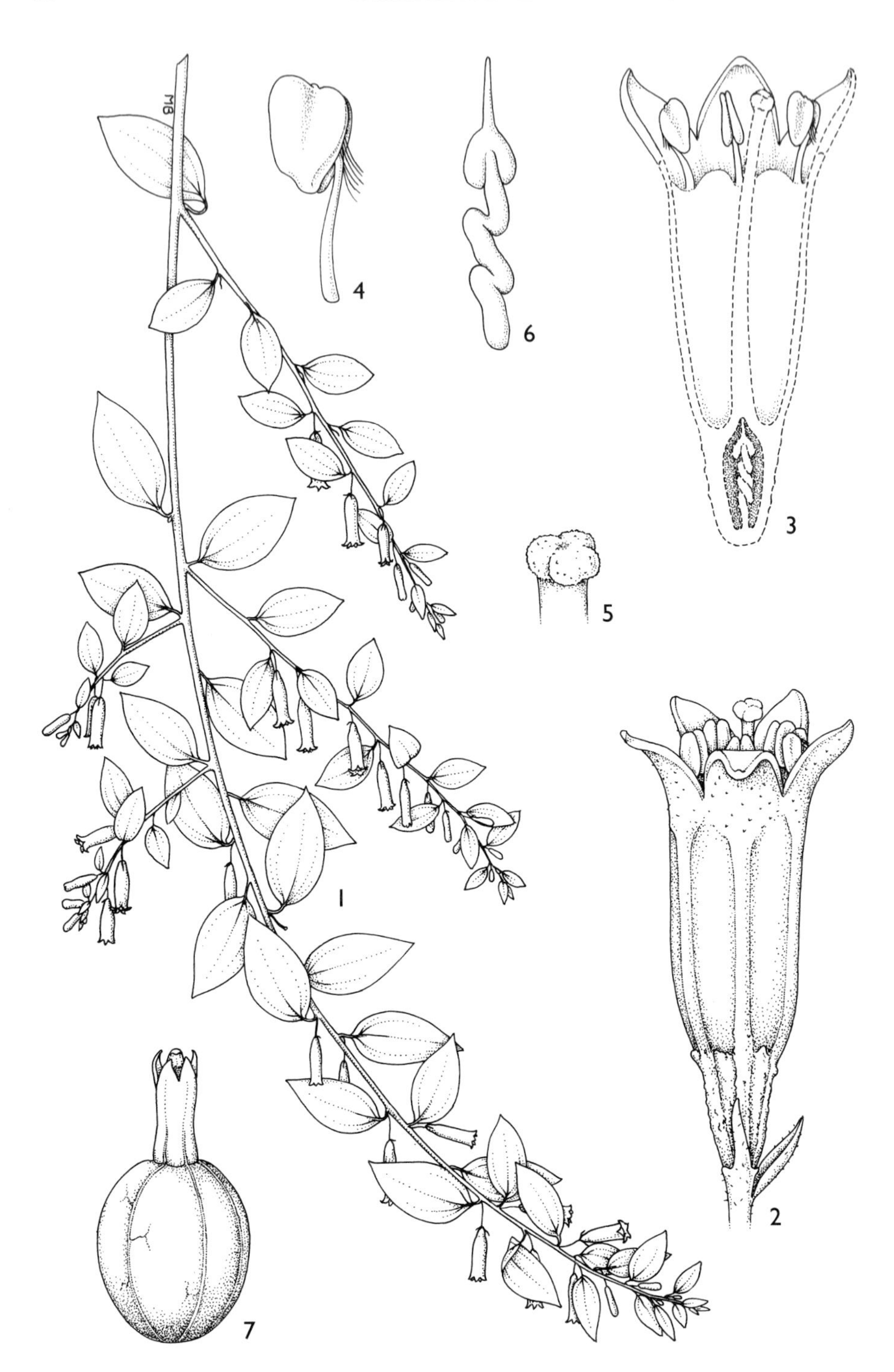

FIG. 4. *OSYRIDICARPOS SCHIMPERIANUS* — **1**, flowering branch, × $^{2}/_{3}$; **2**, flower, × 6; **3**, flower in longitudinal section, × 6; **4**, stamen, × 14; **5**, stigma, × 14; **6**, placenta and ovules, × 14; **7**, fruit, × 3. 1, from *Martineau* 342; 2–6, from *Pawek* 9226; 7, from *Chase* 5049. Drawn by Mary Bates. Reproduced with permission from Flora Zambesiaca.

DISTR. **U** 1; **K** 1–7; **T** 2–5, 7; Congo (Katanga), Eritrea, Ethiopia, Zambia, Malawi, Mozambique, Zimbabwe and South Africa (Limpopo Province, Mpumalanga, KwaZulu-Natal, Eastern Cape Province)

HAB. Upland dry evergreen forest and associated bushland, less often in wetter forests, along rivers and in deciduous woodland; 900–2400 m

SYN. *Thesium schimperianum* A.Rich., Tent. Fl. Abyss. 2: 235 (1850)

Osyridicarpos natalensis A.DC., Prodr. 14: 635 (1857). Lectotype, chosen by Stauffer (1961): South Africa, Eastern Cape Province, King William's Town, Yellowwood R. [Geelhoutrivier], Kachu, *Drège* s.n. (G-DC, holo.; K!, iso.)

O. scandens Engl. in E.J. 19, Beibl. 47: 29 (1894) & P.O.A. C: 167 (1895); Baker & Hill in F.T.A. 6(1): 432 (1911); F.D.-O.A. 2: 143 (1932); Chiovenda, Racc. Bot., Consol. Kenya: 110 (1935); T.T.C.L.: 550 (1949); U.K.W.F.: 336 (1974). Type: Tanzania, Moshi, *Volkens* 1596 (B!, holo.)

O. kirkii Engl. in E.J. 19, Beibl. 47: 30 (1894) & P.O.A. C: 167 (1895); Baker & Hill in F.T.A. 6(1): 432 (1911); F.D.-O.A. 2: 143 (1932); T.T.C.L.: 550 (1949). Lectotype, chosen by Stauffer (1961): Malawi, Chikwawa [Shibisa] to Tshinsunze [Tshinmuzo], Sept. 1859, *Kirk* s.n. (K!, lecto.)

O. mildbraedianus T.C.E.Fries in N.B.G.B. 8: 552 (1923). Type: Kenya, N Nyeri Distrct, Mt Kenya, Liki R., *R.E. & T.C.E. Fries* 1473 (UPS, holo.; K!, iso.)

NOTE. See note under *Thesium triflorum*.

3. **OSYRIS**

L., Sp. Pl.: 1022 (1753) & Gen. Pl., ed. 5: 978 (1754); A. DC. in DC., Prodr. 14: 632 (1857); Stauffer in Viert. Nat. Ges. Zürich 106: 394 (1961); Hilliard in Edinb. J. Bot. 51: 391 (1994)

Colpoon Berg., Descr. Pl. Cap.: 38, t. 1/1 (1767); Stauffer in Viert. Nat. Ges. Zürich 106: 388, fig. 2 (1961)
Fusanus Murray, Syst. Veg., ed. 13: 765 (1774)

Shrubs or small evergreen trees; branchlets ribbed by raised vascular strands running out into midribs of leaves, ultimate twiglets flattened. Leaves petiolate, opposite or alternate, penninerved, coriaceous, both surfaces minutely and closely white-dotted. Flowers hermaphrodite and male (plants androdioecious), axillary and terminal, either solitary or in 2–3-flowered dichasia or umbellate clusters, sometimes panicled; bracts and bracteoles very small, free from peduncle, soon deciduous. Perianth tube very short; lobes usually 3 or 4, valvate. Stamens normally 3 or 4, inserted at base of perianth lobes; anthers attached to lobes by tuft of hairs. Disk fleshy, lobed, lining inside of perianth tube. Ovary with 3–4 ovules pendent from apex of erect free-central placenta; stigma 3–4-lobed. Fruit indehiscent, ovoid, epicarp fleshy.

4 species in Africa and southern Europe to SE Asia.

Osyris lanceolata *Hochst. & Steud.*, Exsicc. Unio Itin. *Schimper* s.n. cum descript. (1832)*; Stauffer in Viert. Nat. Ges. Zürich 106: 388, fig. 1 (1961); K.T.S.L.: 353, fig. (1994); Thulin in Fl. Somalia 2: 145, fig. 94 (1999); M. Coates Palgrave, Trees Southern Africa: 194, photo. 34 (2002). Type: Algeria, *Schimper* s.n. (B!, holo.; B (Herb. Lübeck)!, K!, MO, iso.)

Shrub or small tree, 1.5–9(–14) m tall, all parts glabrous; bark coarsely furrowed; slash bright crimson; branches somewhat flattened and sometimes rather pendent. Leaves usually alternate, elliptic or elliptic-oblong, rarely obovate, 1.5–6.5 × 0.7–4 cm, abruptly apiculate, base cuneate, veins ± immersed, only midvein raised beneath and

* Specimens distributed in year of collection, see Lasègue, Musée Botanique de M. Benjamin Delessert: 121 (1845).

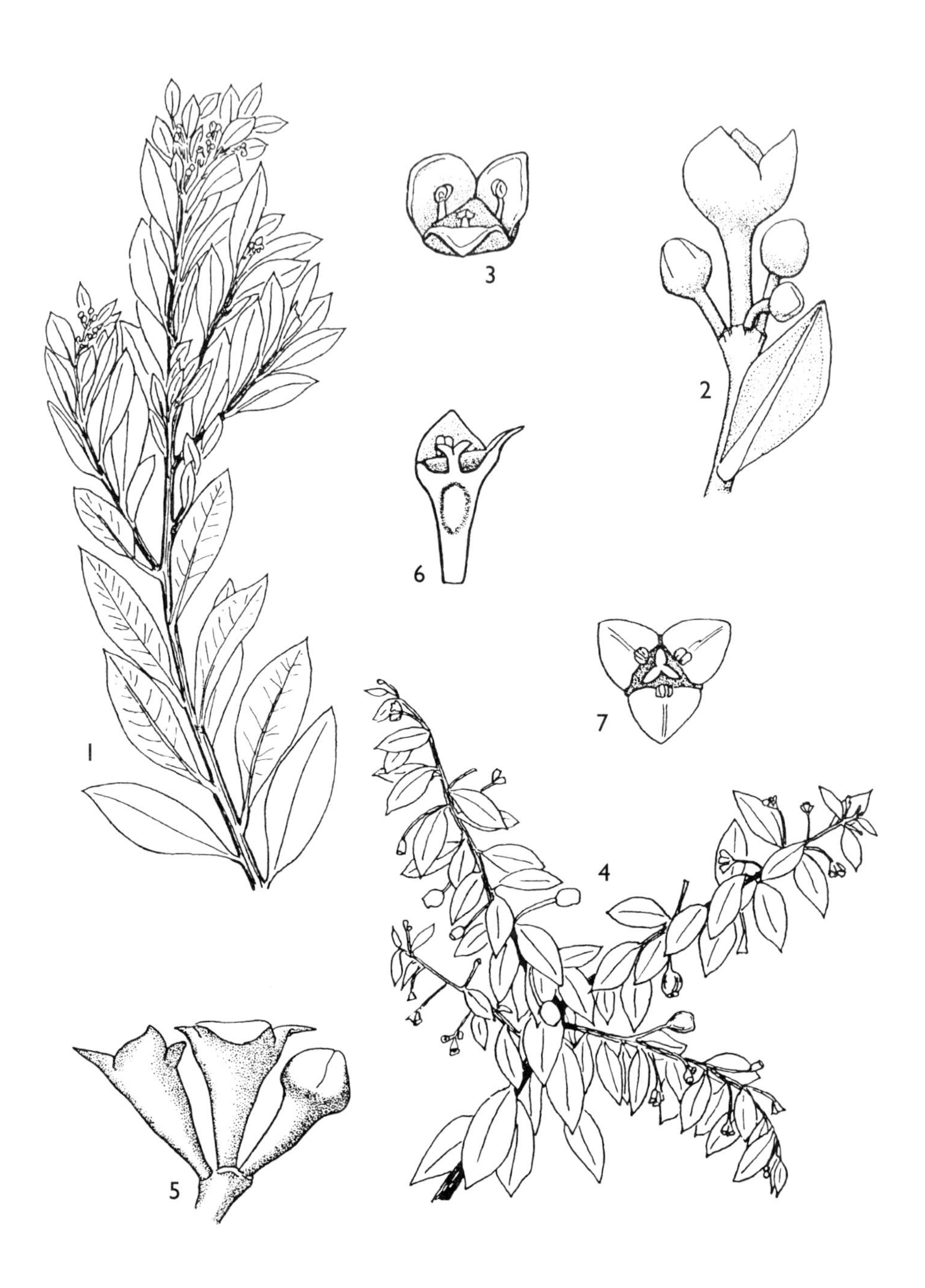

FIG. 5. *OSYRIS LANCEOLATA* — **1**, branch with male flowers, × $^2/_3$; **2**, male inflorescence, × 5; **3**, male flower, × 5; **4**, branch with hermaphrodite flowers, × $^1/_2$; **5**, hermaphrodite inflorescence, × 5; **6**, **7**, hermaphrodite flower, longitudinal section and top view, × 5. Drawn by W.C. Burger and adapted with permission from Burger, Fam. Fl. Pl. Eth., fig. 10 (1967).

running back down stem in a narrow ridge; petiole 1–3 mm long, articulated to a small cushion. Flowers either hermaphrodite or male, plants androdioecious, hermaphrodite flowers in axils of upper leaves; peduncles solitary, usually 1-flowered, occasionally flowers in 2–3-flowered dichasia; male flowers both axillary and terminal, often panicled, each peduncle usually terminating in an umbellate cluster of flowers, rarely flowers solitary or in a 2–3-flowered dichasium; peduncles 4–28 mm long; bracts and bracteoles linear-lanceolate, 1–3 mm long. Perianth yellowish green, leathery; hermaphrodite flowers: tube 0.5–0.6 mm long, obscured inside by disk, lobes 3(–4), spreading, ovate-deltate, 1.5–2 × 1.8–2.2 mm, tips slightly hooded; male flowers similar. Stamens 3(–4); filaments 0.5–0.7 mm long; anthers 0.5 mm long. Ovary 2–2.8 mm long, ovules 3(–4) in hermaphrodite flowers, ovules and placenta aborted in male flowers; style 0.8–1 mm long, thick, cylindric; stigma in hermaphrodite flowers normally 4- rarely 3-lobed, lobes ellipsoid, prominent; in male flowers both style and stigma aborted or rudimentary. Fruit ellipsoid, 5–6.5 mm in diameter when dry, epicarp thin, fleshy, red when ripe. Fig. 5 (page 24).

UGANDA. Karamoja District: Napak, 27 May 1940, *A.S. Thomas* 3588!; Kigezi District: Kamwezi, Feb. 1948, *Purseglove* 2584!; Mbale District: Elgon, Kaburoron [Kaburon], Jan. 1936, *Eggeling* 2485!

KENYA. Northern Frontier Province: Mt Kulal, Gatab, 21 Nov. 1978, *Hepper & Jaeger* 6991!; Kiambu District: Kikuyu Escarpment Forest, 5 km above Lower Lari Forest Guard Post, 13 Dec. 1966, *Perdue & Kibuwa* 8227!; Teita District: Bura, 28 Nov. 1997, *Mwachala* in *Earth watch* 430!

TANZANIA. Mbulu District: Lake Manyara National Park, S of Msasa R., 11 Dec. 1963, *Greenway & Kirrika* 11175!; Ufipa District: Mbisi Forest Reserve, 29 Oct. 1987, *Ruffo & Kisema* 2824!; Songea District: Luhira [Luhila], 21 Oct. 1956, *Semsei* 2612!

DISTR. **U** 1–3; **K** 1–7; **T** 1–8; widespread in Africa from Algeria to Ethiopia and south to South Africa; Europe (Iberian peninsula and Balearic Is.), Asia (India to China), Socotra

HAB. Upland dry evergreen forest and mist forest, with associated bushland and grassland, extending down rivers and from there marginally into deciduous woodland; (50–)900–2700 m

SYN. *O. quadripartita* Decne. in Ann. Sci. Nat., sér. 2, 6: 65 (1836); A.G. Miller in Fl. Eth. 3: 382, fig. 116.3 (1989). Type: Algeria, Tangiers, *Salzmann* (K!, iso.)

O. wightiana J.Graham, Cat. Pl. Bombay: 177 (1839); F.D.-O.A. 2: 145 (1932); T.T.C.L.: 551 (1949). Type: India, Maharashtra, Mahabaleshwar, *Sykes* (CAL, holo.)

O. abyssinica A.Rich., Tent. Fl. Abyss. 2: 236 (1850); Engler, Hochgebirgsfl.: 199 (1892); Baker & Hill in F.T.A. 6(1): 433 (1911); F.D.-O.A. 2: 144 (1932); T.T.C.L.: 551 (1949); I.T.U., ed. 2: 373 (1952). Types: Ethiopia, near Adua, Mt Sholoda, *Schimper* I. 281 (P, syn.; B!, BM!, BR, K!, isosyn.) & Tigray, *Quartin Dillon* s.n. (P. syn.)

O. arborea A.DC., Prodr. 14: 633 (1857); Robyns & Lawalrée in F.C.B. 1: 295 (1948). Types: several specimens from India and Nepal, including Nepal, *Wallich* 4035 (G-DC, syn.; K!, isosyn.)

O. pendula Balf.f. in Proc. Roy. Soc. Edinb. 12: 93 (1884). Type: Yemen, Socotra, Haghier Hills, *B.C.S.* 630 (K!, iso.)

O. rigidissima Engl., Hochgebirgsfl. Trop. Afr.: 199 (1892); Baker & Hill in F.T.A. 6 (1): 433 (1911). Type: Northern Somalia, Surud [Serrut] Mts, Maydh [Meid], *Hildebrandt* 1539 (B!, holo.; BM!, iso.)

O. tenuifolia Engl., P.O.A. C: 167 (1895); Baker & Hill in F.T.A. 6(1): 434 (1911); V.E. 3(1): 69, fig. 38 (1915). Type: Tanzania, Moshi District, below Marangu and Machame [Madschame], *Volkens* 1732 (B†, holo.; BR, iso.)

O. parvifolia Baker in K.B. 1910: 239 (1910); Baker & Hill in F.T.A. 6(1): 434 (1911). Type: Ethiopia, Shoa, Domak, Efat, *Roth* s.n. (K!, holo.)

O. urundiensis De Wild. in Ann. Soc. Sci. Brux. 44, 1: 373 (1925). Type: NE Burundi, region of Lakes Shohoho and Rugweru, collector unspecified (BR, holo.)

O. densifolia Peter, F.D.-O.A. 2: 145, Anh. 11, t. 14/1 (1932); T.T.C.L.: 551 (1949). Type: Tanzania, Masai District, Ol Doinyo Sambu, *Peter* 2142b (B†, holo.)

O. oblanceolata Peter, F.D.-O.A. 2: 145, Anh. 11, t. 14/2 (1932); T.T.C.L.: 551 (1949). Types: Mbulu District, Iraku, Malimo–Dungo, *Peter* 43791 & little waterfalls on Njajede stream, *Peter* 43801 (both B†, syn.)

O. laeta Peter, F.D.-O.A. 2: 145, Anh. 11, t. 15/1 (1932); T.T.C.L.: 551 (1949). Type: Tanzania, Mbulu District, Masai Boma to Kampi ya Faru, *Peter* 43561 (B†, holo.)

[*O. compressa* sensu auct., *non* (Berg.) A.DC.; Brenan in Mem. N.Y. Bot. Gard. 9: 66 (1954); K.T.S.: 500 (1961); White, Dowsett-Lemaire & Chapman, Evergreen For. Fl. Malawi: 520 (2001)]

NOTE. Great variation in leaf size and shape has elicited a considerable synonymy. The above list is not exhaustive. It has at times been included in *O. compressa* (Berg.) A.DC., but that species has the leaves generally opposite, the flowers are all hermaphrodite, the floral parts are generally in fours and the style is much shorter, as elucidated by Stauffer, loc. cit. (1961), and Hilliard in Edinb. J. Bot. 51: 391–392 (1994).

The wood has a fuel value as a substitute for sandalwood according to I.T.U., ed. 2: 373 (1952) & K.T.S.: 500 (1961).

INDEX TO SANTALACEAE

New names validated in this volume

Thesium panganense *Polhill*

First published in 2005 by
Royal Botanic Gardens, Kew
Richmond, Surrey, TW9 3AB, UK
www.kew.org

ISBN 1 84246 113 3

Design by Media Resources, typesetting and page layout by Margaret Newman,
Information Services Department,
Royal Botanic Gardens, Kew.

Printed by Cromwell Press Ltd.

For information or to purchase all Kew titles please visit
www.kewbooks.com or email publishing@kew.org